FROM RURAL VILLAGE
TO GLOBAL VILLAGE

*Telecommunications for Development
in the Information Age*

TELECOMMUNICATIONS
A Series of Volumes Edited
by Christopher H. Sterling

FROM RURAL VILLAGE
TO GLOBAL VILLAGE

*Telecommunications for Development
in the Information Age*

Heather E. Hudson
University of San Francisco

Routledge
Taylor & Francis Group
New York   London

First published by Lawrence Erlbaum Associates, Inc., Publishers
10 Industrial Avenue
Mahwah, New Jersey 07430

Reprinted 2009 by Routledge

Routledge

270 Madison Avenue
New York, NY 10016

2 Park Square, Milton Park
Abingdon, Oxon OX14 4RN, UK

Cover design by Tomai Maridou

Library of Congress Cataloging-in-Publication Data

Hudson, Heather E.
    From rural village to global village : telecommunications for development in the information age / Heather E. Hudson.
        p.   cm. — (Telecommunications)
    Includes bibliographical references and index.
    ISBN 0-8058-5667-6 (cloth : alk. paper)
    ISBN 0-8058-6016-9 (pbk. : alk. paper)
    1. Rural telecommunication.   2. Rural development.   I. Title.
II. Telecommunications.

HE7631.H827   2006
384.3'09173'4—dc22   2005051238
                                                            CIP

10   9   8   7   6   5   4   3   2   1

To the many people in rural and developing regions around the world who have taught me about the importance of communication.

Contents

Preface

This book is intended as a sequel to *When Telephones Reach the Village*, which was published in 1984. In the past two decades, there have been dramatic changes in technology, among them, the advent of the Internet and mobile phones that now outnumber fixed lines in both developing and industrialized regions. As a result, there have also been changes in definitions of access to telecommunications in developing countries, defined in the Maitland Commission report in 1983 as "a telephone within an hour's walk."

This book reviews the research on ICTs (information and communication technologies) and development in the past two decades. Many of these studies have been prompted by technological advances and calls for investment in information infrastructure, information highways, last mile/first mile connections, and so forth. My hope is that readers will find new information and sources that they can pursue to deepen their understanding of this field and to stimulate their own research. The references are by necessity illustrative rather than exhaustive; for example, there are a great many studies on topics such as ICTs in education or telecenters. I have tried to include those that are particularly valuable in their insights or their approach. No doubt I have overlooked some materials that others would consider equally valuable.

The book goes beyond this research, however, to consider policies and strategies to extend access. After participating in many evaluations of the impact of new technologies, I concluded that policy was equally important. Little is gained if the use of ICTs turns out to be very effective in a pilot project, but there is no mechanism to extend the reach of the project or to make its services affordable. Thus access must be coupled with sustainability, topics which are addressed in several chapters of the book.

Readers may be surprised to see references that already appear dated. This is not only because material is always dated by the time it gets into print, but also because the purpose of the book is to examine the transition from rural village to global village. The research, projects, and policy initiatives of the last two decades were all part of this transition. Another reason for including "older" references is that much of what was learned from earlier research remains very relevant today. Lessons from community radio, satellite experiments, and early distance education projects all are relevant for computer and Internet applications. In addition, there is a danger that some of the very valuable studies done during this period will be forgotten, and I hope their inclusion will draw new attention to them.

That said, I have tried to bring the discussion up to date by including recent studies, many of which are available only online. There is a danger that they too could vanish, as Web sites come and go, and links disappear. We owe a debt of gratitude to the universities and other research centers and organizations that have preserved key documents about ICT policy initiatives, task forces, and pilot projects that might otherwise be lost.

I would like to acknowledge the support of many organizations, including the International Development Research Centre, the International Telecommunication Union, the Sloan Foundation, the Ford Foundation, the U.S. Agency for International Development, the Commonwealth of Learning, the Institute for the North, and several others. And most of all, I want to thank the people I have met and worked with throughout the developing South and the remote North, from Nunavut to Namibia and Tanana to Timbuktu.

—*Heather E. Hudson*
Bowen Island, August 2005

The Death of Distance?

The death of distance as a determinant of the cost of communications will probably be the single most important economic force shaping society in the first half of the next century.

—Frances Cairncross (2001)

The end of the 20th century was marked by dramatic technological developments in computers and telecommunications and the emergence of the Internet. It took 75 years before 50 million customers had telephone service, but only four years for the Internet to reach 50 million users. Although these trends are more obvious in industrialized than in developing countries, and generally in urban rather than rural areas, they have enormous promise for developing and rural regions. Technology has effectively erased distance barriers, as members of rural cooperatives in Latin America check futures prices and exchange rates; African teachers and health workers use community telecenters to find information on the Internet, and Asian villagers use cellphones to stay in touch and arrange transport for their produce. Yet this phenomenal change masks both old and new disparities. More than half the world's population has never used a telephone, let alone accessed the Internet. And of those with telephone access, many find it difficult or impossible to use the Internet's World Wide Web because of limited bandwidth or poor service quality.

Access to information is now considered vital to development, so that the classifications "information rich" and "information poor" may mean more than distinctions based on GNP or other traditional development indicators. Information and communication technologies (ICTs) used to pro-

1

duce, store, and transmit information are increasingly important development tools. The purpose of this book is to analyze the role of ICTs for social and economic development in rural areas and developing regions, and then to examine techniques and strategies to close communications gaps, so that the electronic means to access and share information will be available and affordable throughout the developing world.

INFORMATION: THE DEVELOPMENT LINK

Information is critical to the social and economic activities that comprise the development process. Information is obviously central to education, but also to health services, where providers need training as well as advice on diagnosis and treatment of cases beyond their level of expertise or the capacity of local facilities. But information is also critical to economic activities, ranging from agriculture to manufacturing and services.

Distance represents time, in an increasingly time-conscious world. In countries with economies that depend heavily upon agriculture or the extraction of resources (lumber and minerals), distance from urban markets has traditionally been alleviated only with the installation of improved transportation facilities, typically roads. Yet transportation links leave industries without the access to information which is becoming increasingly important for production and marketing of their commodities. Another disadvantage faced by many developing countries is economic specialization. As they strive to diversify their economies, timely access to information becomes even more critical. In the provision of physical goods and services, rural areas could only compete across barriers of distance and geography if they had a natural resource advantage. Now they may need some other characteristic (such as low wages) to compensate for distance from markets, or they may need to specialize (growing exotic fruits, "designer vegetables," shade-grown coffee) and use the power of communications to reach global markets.

Historically, rural development took place in such regions where there was geographic advantage in the form of arable land or natural resources. Increasingly, new economic development depends on human resources, and on economic diversification. Thus, basic education of children and adults as well as specific training are increasingly important. Yet, rural regions worldwide continue to face a shortage of teachers and educational facilities. Typically, rural residents also have much more limited access to health care than their urban counterparts because of a lack of health workers, medical supplies, and clinics. The results are lower literacy levels, lower life expectancy, and higher infant mortality rates in rural than in urban areas. Again, information and communication technologies can help to improve both the availability and quality of education and social services.

Telecommunications is also vital to the emerging information sectors in developing regions. In the provision of information goods and services, reliable telecommunications infrastructure can make geography and distance irrelevant. For example, the National Research Council (1990) points out that for Africa, where populations and economic activities remain largely rural-based, sharing information is vital if Africans are to contribute to finding solutions to their own development problems:

> Economic development in Africa will depend heavily on the development of the information sector. Countries will need the ability to communicate efficiently with local and overseas markets to determine where they many have comparative advantages for supplying their products to consumers or to purchase essential imports, based on current prices and services. Many of the economic development problems facing African countries have scientific and technological components that will require solutions to be developed in Africa by African scientists. . . . Lack of information is a critical constraint.

CHANGING CONTEXTS

Socioeconomic development does not occur in isolation; the analysis of the role of communications in development must be considered in context. Thus it is important to understand the recent changes in many developing economies as well as in telecommunications and information technologies, and the trends that are driving these transitions.

The growth in the service sector is one of the features of the new global economy. Information-based activities account for the largest part of the growth in services, and other sectors are becoming increasingly information intensive. While services tend to be a more visible component of urban economies, this structural shift is mirrored in rural economies of industrialized countries, where public and private services now generally dwarf agriculture and manufacturing. Although this shift may not yet be evident in rural economies of many developing countries, throughout the world, urban and rural activities are being drawn more and more into the global economy. Manufacturers must now be able to respond to changes in demand; suppliers must be able to produce small orders for quick delivery; merchants must be able to update inventory and accounts records instantly. To stay internationally competitive, farmers also must resort to increased specialization, and react to shifts in consumer demand.

The technological context consists of a whirlwind of technological change that tosses up tantalizing but confusing arrays of equipment and services, resulting from breakthroughs in speed of transmission and processing of data and in storage capacity. The transformation from analog to digital communication has made possible convergence of services that were once considered completely separate, including transmission of data, but also voice, music, graphics, and video; the rise of the Internet as a global web of connectivity enables content to be transmitted between individuals or groups or to be "broadcast" (i.e. disseminated to any and all connected to the network). Terrestrial wireless and satellite technologies make possible communication anywhere, anytime. Key technological trends that are driving the proliferation of new information and telecommunications services include:

- **Capacity:** Technologies such as optical fiber have enormous capacity to carry information, and can be used for services ranging from entertainment and distance education to transmission of highly detailed images for remote medical diagnosis. Satellites and some terrestrial wireless technologies also offer a tremendous increase in available of bandwidth.

- **Digitization:** Telecommunications networks are becoming totally digital, so that any type of information, including voice and video, may be sent as a stream of bits. Digital compression allows more efficient use of bandwidth so that customers may have more choices (such as compressed satellite television channels) and/or lower costs, such as use of compressed video for distance education and compressed voice for Internet telephony.

- **Ubiquity:** Advances in wireless technology such as cellular networks, personal communications systems (PCS) and rural radio subscriber systems offer affordable means of reaching rural customers and urban areas without infrastructure in developing countries. Low cost wireless services may also replace wireline in industrialized countries as the primary means of personal access. Wireless broadband technologies can cover not only buildings and campuses but neighborhoods and villages.

- **Convergence:** The convergence of telecommunications, data processing, and video technologies is ushering in the era of multimedia, in which voice, data, and images may be combined according to the needs of users, and distinctions between the traditional sectors of telecommunications, information processing, and broadcasting are increasingly arbitrary and perhaps irrelevant.

There are several significant implications of these technological trends, particularly for rural and developing regions:

- *Distance is no longer a barrier to accessing information.* Technologies are available that can provide interactive voice, data, and multimedia services virtually anywhere.

- *Costs of providing services are declining.* Satellite transmission costs are independent of distance; transmission costs using other technologies have also declined dramatically. Thus communications services can be priced not according to distance, which penalizes rural and remote users, but per unit of information (message, bit) or unit of time.

- *The potential for competition is increasing.* Lower costs make rural/remote areas more attractive. New competitors can offer multiple technological solutions, including wireless, satellite, optical fiber, copper, and cable. Lower costs can translate to lower prices, bringing communication services within the reach of more developing country residents.

In addition, the telecommunications sector is being restructured through privatization of formerly government-owned networks and liberalization to allow competition among services and across technologies. The telecommunications industry itself is also being transformed, with new entrants as well as mergers and acquisitions creating vertically and horizontally integrated providers of both content and conduit, hardware and software. All of these changes are taking place within a context of globalization, as international trade in goods and services expands and national economies become increasingly interdependent.

THE DIGITAL DIVIDE

Of course, these new technologies and services are not available everywhere, and in developing regions where they do exist, many people cannot afford to use them. Telecommunications is a "missing link" in much of the developing world, as the International Telecommunication Union's (ITU) Maitland Commission noted nearly two decades ago. A decade later, policymakers were calling for a "Global Information Infrastructure" that would link everyone into a worldwide network, or more likely, network of networks. By the turn of the century, world leaders were committing themselves to bridge "digital divides" between the industrialized and developing worlds.[1]

Although there has been a dramatic increase in telecommunications investment in the past decade, there are still enormous gaps between the developed and developing world. People in the least developed countries, many of which are in Africa, still have very limited access to basic voice telecommunications. However, wireless technologies are beginning to bridge that gap (see Table 1.1). As will be shown in chapter 8, competition appears as important as technology in accelerating the growth of wireless.

Gaps also exist within the developing world, as telephones are not evenly distributed throughout the population. Not only do wealthier people have greater access to telecommunications, but the gaps are even greater between urban and rural areas. Typically, a high percentage of developing

TABLE 1.1
Access to Telecommunications

Region	Total Telephone Subscribers/ 100 Population	Wireless Subscribers as Percentage of All Subscribers
Africa	6.0	67.5
Americas	61.2	50.3
Asia	20.1	54.0
Europe	84.3	58.9
Oceania	85.0	57.2

Source. International Telecommunication Union, *World Telecommunication Development Report*, 2003. Geneva: International Telecommunication Union.

country residents live in rural areas (as much as 80 percent of the population in the least developed countries), where access to communication networks is much more limited than in urban areas. There are more than three times as many telephone lines per 100 people in the largest city of less developed countries as in the rest of the country. In the poorest countries there may be no access to telecommunications at all in the rural areas. (See Table 1.2; it should be noted that this table overestimates rural access because the "rest of country" includes everywhere except the largest city.)

The term *digital divide* is often used to describe gaps in access to the Internet, both within industrialized countries and between industrialized and developing countries. Table 1.3 shows the gap between the industrialized and developing worlds in three indicators that together suggest the limited access to the Internet. The first is estimated Internet users: More than 85 percent of the world's Internet users are in developed countries, which account for only about 22 percent of the world's population. The other two indicators shown are personal computers and fixed telephone lines because Internet access requires both communications links which are typically dial-up telephone lines in much of the world, and information

TABLE 1.2
Access to Telecommunications

Country Classification	Teledensity (Telephone Lines/100) Population		
	National	Urban	Rest of Country
High Income	46.0	52.9	43.8
Upper Middle Income	13.7	25.7	11.5
Lower Middle Income	9.7	22.1	7.2
Low Income	2.5	6.5	2.3

Source. International Telecommunication Union, *World Telecommunication Development Report*, 2001. Geneva: International Telecommunication Union.

TABLE 1.3
Internet Access Indicators

Region	Internet Users/ 10,000 Population	Main Telephone Lines/100 Population	PCs/100 Population
Africa	156.2	3.0	1.4
Asia	690.7	13.4	4.6
Americas	2644.2	34.2	29.0
Europe	2416.5	41.2	22.4
Oceania	4301.9	40.7	44.9
World	1133.8	18.7	10.1

Source. Derived from ITU data for 2003.

technologies, typically personal computers or networked computer terminals. Although there is still much less access to telecommunications in developing countries than in industrialized countries, at present, the gap in access to computers is much greater than the gap in access to telephone lines or telephones. High-income countries had 22 times as many telephone lines per 100 population as low-income countries, but 96 times as many computers. Despite the falling cost of computers, they are still out of reach of much of the world's population. A computer costs the equivalent of 8 years' income for the average person in Bangladesh, but less than a month's wages for the average American.

It should be noted that broadband access is not included in Table 1.3. Recent data from the ITU indicates that some broadband services were commercially available in 82 economies (out of 200) worldwide, but that penetration rates are quite closely correlated with gross national income per capita, so that in developing countries that do have broadband, access is generally very limited.[2] Yet there is cause for optimism. New technologies offer the possibility of technological leapfrogging, for example, to reach end users through wireless local loops or small satellite terminals rather than stringing wire and cable. Digital transmission and switching are increasing reliability and lowering cost, as well as making it possible for subscribers in developing countries to use electronic mail and voice messaging, and to access the Internet.

Yet while access is necessary, it is not sufficient to ensure that disadvantaged populations benefit from ICTs. In its Statement on Universal Access to Basic Communication and Information Services, the United Nations noted:

> The information and technology gap and related inequities between industrialized and developing nations are widening: a new type of poverty—information poverty—looms. Most developing countries, especially the Least Developed Countries (LDCs) are not sharing in the communications revolution, since they lack:

- affordable access to core information resources, cutting-edge technology and to sophisticated telecommunications systems and infrastructure;
- the capacity to build, operate, manage, and service the technologies involved;
- policies that promote equitable public participation in the information society as both producers and consumers of information and knowledge; and a work force trained to develop, maintain and provide the value-added products and services required by the information economy.[3]

These goals of affordable access to ICTs, capacity to use and manage them, and policies that promote these goals are a major focus of this book.

FROM RURAL VILLAGE TO GLOBAL VILLAGE

The study of the impact of communication technologies on development focused almost exclusively on mass media until the 1980s. Whereas interest in educational radio in the industrialized world declined with the growing popularity of television, radio in the developing world continued to offer an inexpensive (for both producer and listener) means of transmitting information and educational content. Development agencies such as UNESCO and USAID funded radio projects for formal education, either to teach children directly or to upgrade teachers' knowledge and skills, and for nonformal education, to reach youths not in school and adults at home and at work. As television became available in developing countries, its educational potential was explored and evaluated in early projects in Samoa, El Salvador, and Senegal, and eventually in many more developing countries. Yet two-way or interactive communications were generally ignored, except by a few pioneering technical scholars such as Colin Cherry (1957, 1971), who sought to understand the social and economic impacts of telecommunications in the industrialized world. During the mid 1970s, studies on interactive applications such as telemedicine and teleconferencing began to appear, many of which were funded as part of technical experiments using new technologies such as satellites (see e.g., Hudson & Parker, 1973).

Beginning in the late 1970s, research on the impact, or potential impact, of interactive telecommunications began to appear, primarily instigated by the International Telecommunication Union (ITU) and the World Bank. Much of the early research was synthesized in one of the first comprehensive reviews, *When Telephones Reach the Village* (Hudson, 1984). This book is intended to update and expand upon that earlier study. The purpose of this book is to present what we have learned in the interim about how new information and communication technologies and services can contribute to social and economic development, and to propose policies and strategies so that they are truly accessible (i.e., both available and affordable) throughout rural and developing regions. The book reviews key studies and identifies theory and research findings concerning rural regions and developing countries. The primary emphasis is on developing countries, but research conducted in industrialized countries is included if it appears particularly relevant for rural regions of the developing world. The analysis is accompanied by an extensive bibliography, which is by design illustrative rather than exhaustive, as the literature on information and communication technologies (ICTs) in some sectors such as education and health care is very extensive.

The first section examines findings from macrolevel studies on the role of telecommunications in development, and then summarizes findings from studies of rural sectors such as agriculture, fisheries, natural resources, crafts, and small and rural businesses. The following chapters focus on the role of ICTs in education and training and on health care delivery. The next section of the book addresses telecommunications technologies that hold promise for reaching unserved populations, an analysis of the status and evolution of the "digital divide," and policy issues that must be addressed if access to information and telecommunications networks is to be universally available and affordable. Several case studies are included to provide practical examples of the issues addressed in the book: on telecenters for community access to the Internet; telehealth in Alaska and Africa; distance education in northern Canada, the West Indies, and the South Pacific; and the proliferation of mobile phones in developing regions.

NOTES

1. Okinawa Charter on the Global Information Society. See www.g8.utoronto.ca/summit/2000okinawa/gis/html
2. See *ITU Internet Reports: Birth of Broadband. Executive Summary.* Geneva, ITU, September 2003. There are some notable exceptions of emerging economies with unusually high broadband penetration, such as Estonia and South Korea.
3. United Nations Administrative Committee on Coordination (ACC), "Statement on Universal Access to Basic Communication and Information Services," April 1997. Quoted in ITU: *World Telecommunication Development Report,* 1998, p. 10.

Telecommunications
and Rural Development

We have a saying: "When the telephone rings, business is coming."
—Manager of a rural cooperative in China

In the past two decades, there has been growing interest in the role of telecommunications in the development process, sparked perhaps by the heightened visibility of the sector, as technological change and competition have fostered new facilities and services. The advent of the Internet has focused new attention on the properties of information and information networks. Researchers with technical, economic, and other social science backgrounds have attempted to understand and quantify the role of information in economies. They have been joined by rural development specialists who are asking whether investment in information infrastructure can contribute to rural economic growth and diversification.

This chapter summarizes findings from research on applications of ICTs for rural development employing aggregate data analysis, field research, and case studies. Topics include agriculture, fisheries, crafts, and other entrepreneurial activities. It then examines profiles of users, and impacts specifically related to women, and complementary factors that are required for benefits of ICTs to be optimized.

THE POWER OF INFORMATION SHARING

Information (or knowledge, which we may consider to be an organized body of information) has unusual properties in that it may be shared without being transferred, and its benefits may extend to those who have not directly participated in its acquisition, processing, or dissemination. For example, if a physician teaches health workers how to treat a disease, all now have that knowledge; the health workers may then be able to treat patients who benefit from that sharing of knowledge. These indirect benefits are also known as externalities.

Information networks have two types of externalities. Subscriber-related externalities are benefits and costs that accrue to existing subscribers from expansion of the system; they accrue to all subscribers. Each new telephone subscriber adds more value to the others in the network; each new Internet user and website adds new sources of information for all. Call- or message-related externalities are benefits and costs that accrue to the caller or message sender and anyone contacted. Thus, in a telecommunications network, both the caller and person called may benefit from the information shared, although typically only one of them pays for the call. Third parties may also benefit; for example, a farmer would benefit if an agricultural extension agent could contact an agronomist to find out how to eliminate a crop fungus. Similarly, a patient would benefit if a doctor were able to find out how to treat her rare disease by searching on the Internet. Some benefits may accrue to individuals who use ICTs, such as getting help in emergencies by contacting a doctor, and saving time by using telecommunications to arrange transport logistics or to substitute for traveling to the city. Other benefits may require more complex types of information seeking or use by people with institutional affiliations, such as members of community organizations, entrepreneurs, employees of nongovernmental organizations (NGOs), businesses or government agencies.

Information networks such as the Internet magnify the power of information sharing; a property of information networks is that each subscriber's welfare rises with the number of subscribers who have access to the network. A basic principle of connectivity known as Metcalfe's Law (coined by Robert Metcalfe, inventor of Ethernet and an Internet pioneer) is that the number of connections and thus the potential value of network increases almost as the square of its users (quoted in Shapiro & Varian, 1998, p. 184). The more people have e-mail accounts, the more valuable e-mail services are to each user. Coupling e-mail with the potentially infinite number of websites as information sources, the Internet has enormous potential value to facilitate the sharing of ideas and through this communication, the expansion of knowledge. This principle is critical for telecommunications policy, because it implies that building out the network and connecting more subscribers should be the most important priority for telecommunications planners and policymakers, as discussed in chapters 5 to 8 on policy.

THE BENEFITS OF CONNECTIVITY

Information is critical to the social and economic activities that comprise the development process. If information is critical to development, then ICTs, as means of sharing information, are not simply a connection between people, but a link in the chain of the development process itself. In general, the ability to access and share information can contribute to the development process by improving:

- *Efficiency,* or the ratio of output to cost (for example, through use of just-in-time manufacturing and inventory systems, through use of information on weather and soil content to improve agricultural yields);
- *Effectiveness,* or the quality of products and services (such as improving health care through telemedicine);
- *Equity,* or the distribution of development benefits throughout the society (such as to rural and remote areas, to minorities and disabled populations);
- *Reach,* or the ability to contact new customers or clients (for example, craftspeople reaching global markets on the Internet; educators reaching students at work or at home).

Examples of the contribution of telecommunications to *effectiveness* include distance education, where real-time instruction and tutorials are more effective than simply requiring students to study correspondence materials, and telemedicine, where information about the patient transmitted electronically can help distant specialists to diagnose conditions and recommend treatment. ICTs contribute to *equity* by enabling the disadvantaged, including the poor, isolated rural people, and the disabled, to obtain information that would otherwise be very difficult or impossible to access.

Some researchers have attempted to summarize the benefits of telecommunications investment for various sectoral activities. Saunders, Warford, and Wellenius (1994) group benefits under four categories:

- Market information for buying and selling;
- Transport efficiency and regional development;
- Isolation and emergency security;
- Coordination of international activity, including business, tourism and international organization.

More specifically, among the benefits of ICTs for improving efficiency and productivity are the following:

- *Price information:* Producers such as farmers and fishermen can compare prices in various markets, allowing them to get the highest prices for their produce, to eliminate dependency on local middlemen, and/or to modify their products (types of crops raised or fish caught, etc.) to respond to market demand.

- *Reduction of downtime:* Timely ordering of spare parts and immediate contact with technicians can reduce time lost due to broken machinery such as pumps, tractors, and generators.

- *Reduction of inventory:* Businesses can reduce the inventories they need to keep on hand, and therefore the capital they would be tied up in inventory, if replacements can be ordered and delivered as needed.

- *Timely delivery of products to market:* Contact between producers and shippers to arrange scheduling for delivery of products to market can result in reduced spoilage (for example, of fish or fresh fruit), more efficient processing and higher prices for produce.

- *Reduction of travel costs:* In some circumstances, communications may be substituted for travel, resulting in significant savings in personnel time and travel costs.

- *Energy savings:* Telecommunications can be used to coordinate logistics of shipping and transportation so that trips are not wasted and consumption of fuel is minimized.

- *Decentralization:* Availability of information networks can help to attract industries to rural areas, and allow decentralization of economic activities away from major urban areas (Hudson, 1997).

The following are examples of these benefits for rural sector activities.

AGRICULTURE AND FISHERIES

As noted earlier, timely access to information may be particularly useful for farmers and fishermen to improve yields and get better prices than available through middlemen. In Chile, an Internet network linking farmer organizations has dramatically increased farmers' income by providing information about crop status, global market prices, and training.[1] In Pondicherry, India, the MS Swaminathan Research Foundation has set up rural information centers that enable farmers to access information such as market prices, so that they can negotiate better with intermediaries, while fishermen can download satellite images that indicate where to find fish shoals. About one third of the users are from assetless households and about 18 percent are women.[2]

Brazilian farmers used newly installed telephones to contact the Chicago futures exchange, to get futures prices for coffee. Based on that information, they could decide whether to sell their crop immediately or hold it to obtain a better price in a few months. Now they could get this information online. Farmers in the Nile Delta call merchants in Alexandria to get orders directly, rather than selling at lower prices to local middlemen (Hudson, 1992). Similarly, Sri Lankan farmers were able to negotiate higher prices after obtaining market information from Colombo using local payphones (Saunders et al., 1994). Today, cell phones can serve the same purpose. In fact, a provider of cellphone service in Uganda offers daily price information from the markets in Kampala for coffee, vegetables, and fish. Subscribers can access that information for the price of one text message.[3]

Logistics can also be important in agriculture and fisheries, where perishable goods must reach the market before they spoil. Filipino fishermen now use cell phones to book cargo space on aircraft that fly freshly caught tuna to Japan, where world prices are highest. In northern Canada, indigenous lake fishermen use two-way radios to alert float plane operators when their catch is ready to be flown to urban markets. In the past, prearranged flights were wasted if the catch was too small or the fish were no longer fresh. Cook Islanders used two-way radio to adjust the itinerary of ships dispatched to pick up fresh fruit destined for New Zealand. Otherwise, if ships did not arrive on time, the fruit would be overripe, and would fetch a much lower price.

EXTENDING REACH: GLOBAL MARKETS

The Internet offers crafts people from rural regions around the world a shopfront in the global marketplace. For example, through PEOPLink's global artisans trading exchange, local craftspeople are increasing their income through access to new markets, effectively removing wholesale intermediaries. They can now receive up to 95 percent of the selling price for their produce whereas previously they received only 10 percent.[4] In Kenya, the Naushad Trading Company, which sells African handicrafts, woodcarvings, pottery, and baskets, saw revenue grow from US$10,000 to over US$2 million in its first two years.[5] Indigenous artisans in remote areas of North America and Australia are also reaching global markets via the Internet. The Oomingmak Cooperative in Alaska showcases on its website the clothing knit by native Alaskan women from the fine wool of the muskox, said to be the warmest in the world. Inuit from the Canadian Arctic display highly valued soapstone carvings on the Internet, while aborigines in Australia market their paintings and digeridoos on the Web.

Remittances are a significant source of income in many developing countries, as workers abroad send money home to their families. It is estimated that global remittances from migratory workers now exceed $100 billion per year, with 60% going to developing countries.[6] An estimated $30 billion in remittances was sent from the United States to Latin America in 2004, with more than 60 percent of Latin American born residents in the United States sending money home on a regular basis.[7] These funds are transferred primarily through banks and companies such as Western Union. However, developing country entrepreneurs have also found that they can use telecommunications to provide services funded by remittances. An Ethiopian Web site offers to deliver gifts including flowers, cakes, and sheep within the country. Overseas Ethiopians may pay with credit cards.[8] A Peruvian e-commerce company realized after the Internet bubble burst, that Peruvians had little interest in online shopping and reinvented itself as a site for remittance-based businesses. Migrants can go online and pay remittances to their family members, who can use a Peruvian Visa card to make purchases in Peru.[9]

Microfinance may be a valuable form of e-commerce in developing countries. For Mexico, for example, with more than $13 billion per year in transfers from the United States, Robinson proposes establishment of a rural microfinance system combining microbanks and telecenters. He estimates that typical transfer and communication costs take about 20 percent of this transfer, but that such a microfinance system could provide the services at a much lower rate, and could be combined with other services such as Internet access and Internet telephony. "The operational premise here is that practical people will use the system if it represents a significant savings over current practices and if it is secure and available close to home on both ends. . . . The goal is to develop a managed and secure system that drops the transfer cost to a minimum, estimated to be circa 5 percent, while providing a series of Internet technology-based benefits and services."[10]

MEASURING ECONOMIC BENEFITS

Some researchers have attempted to identify and quantify the externalities associated with telecommunications utilization. An analysis of the welfare effects of investment in telecommunications facilities in developing countries noted that lower transaction costs and reduced uncertainty can increase the efficiency of both markets and administrative organizations, and concluded that telecommunications projects can produce significant external economies, and that the welfare consequences of telecommunications expansion include public-good effects, "notably in a country's capacity for responding to new problems and opportunities" (Leff, 1984). Empirical research and case studies have shed considerable light on the ways and extent to which ICTs contribute to development. Studies sponsored by agencies such as the International Telecommunication Union (ITU) and the World Bank have shown that telecommunications can facilitate many development activities including agriculture, fisheries, commerce, tourism, shipping, education, health care, and social services. These studies show ratios of benefits to costs of telecommunications usage from 5:1 to more than 100:1 from improved efficiency in managing of rural enterprises, time savings in ordering spare parts, savings in travel costs and time, etc.[11]

In the 1970s, several studies noted a high correlation between economic growth and telecommunications investment, typically measured in GDP per capita, and telephone sets or lines per 100 population. However, these studies did not answer the chicken-and-egg question: Does telecommunications investment contribute to economic growth, and/or does economic growth result in increased telecommunications investment? The first major study to address the causality issue underlying the correlation between telecommunications investment and economic growth was by Hardy, who analyzed data from 60 nations over a 13-year period using path analysis and cross-lagged correlation techniques, with time-lagged offsets on one year. Hardy found that the causal relationship ran in both directions. Of course, telecommunications investment did increase as economies grew, but there was also a small but significant contribution of telecommunications to economic development. The implication was that early investment in telecommunications could contribute to economic growth. A causal relationship did not hold for radios per 1000 as a contributor to per capita GDP growth; it was therefore postulated that the organizational impact of telecommunications contributed to economic growth. The study also found that the economic impact of adding telephone lines is comparatively greater in regions with low telephone density, which are likely to be rural and remote areas (Hardy, 1980).

Models such as those developed by Hardy (1980) and Hansen et al. (1990) as well as field research, indicate that economic benefits of telecommunications are related to distance and density, so that benefits are proportionately greater where telephone density is low and alternatives to communicate are expensive and/or time consuming. Also, where telecommunications services are available, rural people often use them more heavily and spend more of their disposable income on telecommunications than do city dwellers. Telecommunications operators frequently find that demand in rural and remote areas is greater than forecasts based on population alone would indicate. In northern Canada, Indians and Inuit (Eskimos) spend more than three times as much as their urban counterparts on long distance telephone calls, even though their average income is generally much lower than that of urban Canadians. The number of long distance calls in some villages in northern Canada increased by as much as 800 percent after satellite earth stations replaced high frequency radios. In Alaska, the installation of small satellite earth stations in villages also sparked tremendous growth in telephone use. When local telephone exchanges were installed in some villages, long distance telephone traffic spurted again by up to 350 percent (Hudson, 1990).

In developing countries, Hardy found that residential telephones appear to contribute more to economic development than business telephones (Hardy, 1980). The reason may be that residential phones are often used for business activities, and are available 24 hours per day, whereas business phones are available only during work hours. There may also be a difference between public and private sector use, with many businesses using their proprietors' home telephones (Hardy, 1980). Now cellphones, which are the first and only telephones for many in the developing world, are likely to serve both personal and business or organizational purposes (see chapter 7).

Hudson, Hardy, and Parker (1982) applied the Hardy methodology to estimate the economic impact of installing small satellite earth stations in three groups of developing countries and a hypothetical rural region. A logarithmic relationship between telephone density and impact on GDP per earth station was found, so that impact per earth station increased with lower telephone densities. They suggested that the model could be used to estimate the impact on national GDP of installing telephone lines and/or thin route satellite stations in regions of low telephone density. Parker (1983) also applied these findings to estimate of the U.S. REA (Rural Electrification Administration) telephone loan program, which provides low-cost loans to rural telephone companies to install and upgrade their networks. Estimating the economic value per telephone line based on Hardy's findings, he calculated the consequent reduction to U.S. GDP if these lines had not been installed. He estimated that the benefits were 6 to 7 times higher than costs to government in interest subsidies.

Cronin, Parker, Colleran, and Gold (1991) used a similar methodology to address the question: "How does the strong relationship between telecommunications and economic development occur?" Their analysis of 31 years of U.S. data shows not only that increases in output or GNP level lead to increases in investment in telecommunications, but that the converse is also true: increases in telecommunications investment stimulate overall economic growth. Another study by Cronin, Colleran, Herbert, and Lewitzky (1993) investigated how telecommunications contributes to national productivity; it showed that incorporation of telecommunication technology and services into production processes has a statistically significant impact in almost all industries. The analysis includes manufacturing, private non-farm, and total private business sectors. The findings suggest that investment in telecommunications infrastructure is causally related to the national total factor productivity and that contributions to aggregate and sector productivity growth rates from telecommunications advancements are both quantifiable and substantial. They conclude that since 1978, about 25 percent of total direct and indirect aggregate productivity gains in the U.S. economy resulted from advances in telecommunications production and enhanced consumption possibilities for end-user industries. A recent study by Waverman, Meschi, and Fuss (2005) finds that mobile phones have a significant impact on economic growth in developing countries.

Aside from the few macrolevel or longitudinal studies, case studies provide much of the evidence on the benefits of telecommunications in rural development. For example, some sector-related benefits discussed below are derived from findings of ITU-supported case studies on the role of telecommunications in agriculture, fisheries, transport, marketing, and other economic activities.[12] However, methodologies vary, making it difficult to extract quantified benefits that could be generalized to other contexts.

Some field studies have analyzed the purposes of telephone calls and time savings compared to other methods of communicating. Kaul (1978) used estimates of costs of transportation alternatives and time lost to estimate the value of telephone calls in rural India. Mayo, Heald, and Klees (1992) used similar methodologies to estimate the value of telephone calls made over a rural satellite network in Peru. They note that businesses generated about one third of the calls made over the network, and that business users estimated that each call saved them about $7.30 compared to alternative means of communication. A survey of rural telephone users in Indonesia and Thailand found that residents of villages and semirural towns had pressing needs to communicate with people outside their own surroundings to perform work-related duties, conduct business transactions, deal with government offices, and reach family members working away from home. The available telephones were used extensively; three quarters of calls were long distance. The alternative was to travel long distances to deliver messages (Chu, Srivisal, & Supadhiloke, 1985).

SOCIAL BENEFITS OF USING ICTs

Many researchers note that there are obvious social benefits of telecommunications, particularly in rural and isolated areas where other forms of interaction with distant family members, friends, and colleagues may be infrequent and time consuming. The most comprehensive review of these findings is the collection edited by Pool (1977), although its focus is on industrialized countries. Field research from developing countries cites examples of rural residents keeping in touch with family members who have gone to the city or overseas to seek work (such as Egyptians and Indians to the Arab Peninsula, or South Pacific islanders to New Zealand); families contacting relatives scattered in many rural communities; and field staff such as nurses and teachers in rural posts using two-way radios or satellite links to keep in touch with colleagues and family members. Moyal (1992) describes the social role of the telephone among isolated families in the Australian Outback: "The role of the telephone [for Australian women] has hence changed . . . from an important facility for expediting daily life and transforming the problem of distance, to an arena where the claims of feeling, care giving, and social support are explicitly acknowledged" (p. 58). Some researchers refer to social calls among friends and family as "intrinsic" use, as opposed to "instrumental" use for business purposes.

Researchers have hypothesized that reducing isolation can help to reduce personnel turnover. Although causal data are difficult to obtain, it appears that communication is at least an important factor. For example, better communications is cited as one of several factors encouraging reversal of the medical braindrain in Navrongo, Ghana.[13] There is anecdotal evidence from northern Canada, Alaska, and Outback Australia that the ability to stay in touch with family and friends makes isolated postings more tolerable, and may contribute to reducing turnover among field staff. Urban-trained field staff from developing countries have expressed similar views on the importance of telecommunications in coping with isolation (see Hudson, 1990, 1992). It appears that telecommunications is one of several factors that may tend to reduce staff turnover, with other benefits such as pay bonuses, travel, and continuing education also being important.

CHARACTERISTICS OF USERS AND "INFOMEDIARIES"

People in developing regions who need information for their work include entrepreneurs, project managers, educators, and health care providers. Individuals may also need to communicate for personal reasons, to stay in touch with family and friends who have gone away to work, as well as for personal business. In Egypt, it was found that better educated individuals were more likely to make calls to major cities and administrative centers, whereas those with little education tended to call only to nearby villages and towns (Pierce & Jequier, 1983). A Costa Rican study of rural public call offices (PCOs) found that villages that benefitted most from the PCOs tended to be larger and better off economically, with relatively better educated populations engaged in more progressive agricultural techniques. The PCO users themselves tended to be employed, but their incomes were not higher than average. In fact, telephone users included a substantial number of low-income residents, although the most frequent callers had higher than average incomes (Saunders et al., 1994). In the United States,

Schmandt, Williams, Wilson, and Strover (1990) suggested that large farmers and agribusiness benefit most from online agriculture data. World Bank reports cite several studies examining characteristics of telephone users and purposes of use.

However, the most important characteristic of ICT users is thirst for information. For example, village chiefs in northern Canada without formal education may use the telephone to talk to other chiefs, and villagers who do not speak the national language or have limited education may rely on intermediaries to obtain information, such as an extension agent, cooperative manager, or other official who in turn will use the telephone to obtain the information they need. Thus, although telephone users tend to be better educated and more involved in the market economy than nonusers, literacy and "modernity" are not prerequisites for telecommunications use. Information seekers may be traditional people concerned about their families, their work, or problems in their community. They are likely to use whatever tools are available—from two-way radios to e-mail—to find the information they need (Hudson, 1992).

In developing regions, the need for services besides basic voice is now spreading beyond urban areas, businesses, and organizations, to small entrepreneurs, NGOs (nongovernmental organizations), and students, driven by demand for access to e-mail and the Internet. E-mail is growing in popularity because it is much faster than the postal service and cheaper than facsimile transmission or telephone calls. Literate young people are flocking to cybercafés and community telecenters in the developing world to use these new tools.

Of course, literacy is required to benefit from access to e-mail and most information available on the Internet. Yet such services can be valuable even for illiterates. When asked about the potential benefits of the Internet for his people, a member of parliament from Uganda stated: "My father sent many telegrams during his life. My father could not read or write." He was able to get a literate person to write his messages down and read the responses to him. Today, his father could get help to communicate via e-mail as well. Similarly, villagers who do not speak the national language or have limited education may rely on intermediaries such as an extension agents, health workers, librarians, cooperative managers, or other skilled users to obtain the information they need from the Internet or to send e-mail and text messages.

This model of getting relevant information through an intermediary could make the benefits of the Internet widely available in developing regions with low literacy levels. In fact, many ICT projects are designed to include intermediaries, sometimes called "infomediaries" or "information brokers," staff members or volunteers who can assist others in finding information. Illiterate farmers in Ecuador were able to get advice on how to eradicate a pest that was destroying their potato crop from an ICT field worker who posted a question about their problem on several e-mail news groups. The responses to her queries suggested a solution that saved the farmers' crop and livelihoods.

GENDER ISSUES: WOMEN AS USERS
AND BENEFICIARIES

In many developing regions, women are using ICTs to obtain and share information relevant to their work, as teachers, community development workers, artisans, or entrepreneurs. These women may benefit directly from ICTs, but many more women may benefit indirectly if information is obtained that is relevant for their work or family, or if an intermediary tracks down the information on their behalf. For example, in many parts of the developing world, women do much of the agricultural work. They may also take the crops to market, or negotiate the price for their crops or livestock with merchants. In such cases, the benefits of telecommunications in getting information about prices and markets, and getting expert advice from extension agents should apply to women. Mass media, such as farm radio forums, which were developed in Canada, and then adopted in India and other developing countries, have reached women as well as men with agricultural information (cf. Atkins, 1990).

Concerning telephone use, Meerbach (1991) suggests that some rural women were using public telephones in Senegal, but that access means more than availability of a telephone in the village. Women may not have identified business contacts to call, may not be able to afford to call, and may not even know how to place a telephone call. Meerbach emphasizes that before implementing a telecommunications system, it is imperative to understand the economic, social, and psychological background of the target group, as well the power relationships within that setting.

A study by Lebeau (1990) that examined payphone use in four rural villages in Botswana found that the vast majority of payphone users (72 percent) were women. The report concluded that telecommunications provision can affect men and women differently, and that usage patterns differed between male and female. For the four Botswana villages, women made and received more calls than men, but spent less per call. Men were more likely to choose an alternative to payphone use that involved travel, and was more costly. On average, female telephone users were younger than males, and appeared to have different communities of interest.

The authors point out that this research was based on secondary analysis, which limited the study and the implications that could be drawn from it on benefits to women and constraints, such as access to transportation. They suggest that "gender sensitive" research techniques be used in the future, such as inclusion of more detailed categories on purpose of call, realistic alternatives to phone use, occupation categories that apply to women, etc. Although studies on direct benefits to women are few, it appears that women do benefit either as participants (teachers, health care workers, farmers, etc.) or indirectly through information that benefits them as mothers, entrepreneurs, employees, or community residents.

Women entrepreneurs are involved in reselling cell phone service in countries such as Bangladesh, the Philippines, and Mozambique. They may physically carry cellphones on foot or bicycle, acting as portable payphones, or may sell minutes on phone cards, or even transfer minutes electronically to users' phones in the case of the Philippines. Women entrepreneurs also manage phone shops in Ghana and teleboutiques in Senegal and Morocco (Hafkin & Taggert, 2001), and many community telecenter staff in Africa and other developing regions are women.

Concerning the Internet, Hafkin and Taggert (2001) note that it is very difficult to get reliable statistics on women's use of the Internet in developing countries. They assert that most women Internet users are not representative of women as a whole, but of a small, urban educated elite, which is true of Internet use in general. Their data show that 22 percent of all Internet users in Asia, 38 percent in Latin America, and 6 percent in the Middle East are women. No data by gender were available for Africa. Despite this paucity of data on Internet usage, it is possible to apply the same analytical framework as for women using telephones—they may participate directly as users, or may benefit indirectly from the information available through ICTs.

In some developing countries that have attracted outsourced information services, women are now employed in information processing and call centers. Early leaders were countries with English speaking workers such as the West Indies and the Philippines; India is now the leader in offshore customer support. Yet these are primarily urban jobs for educated workers. Few women are employed in more highly skilled IT jobs, but in countries that produce and assemble IT components, largely in Asia but also in Mexico and some other parts of Latin America, women are employed in assembly jobs. However, few of these jobs have been created in Africa.

Women are indirect beneficiaries of ICT use in a variety of contexts. Improvements in health care (see chapter 4) can benefit women and their families; use of ICTs in schools (see chapter 3) can improve education for girls, and availability of ICTs in the community can provide opportunities for training as well as tools that can be used by community organizations that benefit women. But, as noted below, availability of ICTs alone will not guarantee developmental benefits.

ICTs AS COMPLEMENT: NECESSARY
BUT NOT SUFFICIENT . . .

ICTs can contribute significantly to socioeconomic development, but investments in them alone are not sufficient for development to occur. Thus ICTs may be seen as a complement to other infrastructure required for development such as transportation, electrification, and a clean water supply. Effectively managed operations may also benefit more from ICT investment. For example, a well managed decentralized organization such as a manufacturing enterprise, a tourist development, or a health service will derive more benefits from ICTs than a poorly managed or understaffed operation. To summarize, ICTs are a necessary but not sufficient condition for economic development (Parker & Hudson, 1995; Schmandt et al., 1990).

ICTs may also serve as a catalyst at certain stages of the development process, becoming particularly important when other innovations are introduced such as improved farming practices, lines of credit, incentives for decentralization, and diversification of the rural economic base. Evidence from India suggests that telecommunications becomes more important when rural modernization begins, for example, when improved farming practices are introduced or credit is made available, or when an integrated development plan is being implemented. In *Electronic Byways*, Parker and Hudson (1995) cite examples such as the tax incentives South Dakota used to lure Citibank's credit card operations from New York, the low wages and tax structures of some Midwest states that have fostered the growth of telemarketing services there, and pollution controls for Los Angeles that are likely to stimulate the growth of telecommuting.

In analyzing European Community rural telecommunications projects, Martin and McKeown (1993) state that ". . . neither EC intervention nor application of ITT (Information and Telecommunications Technologies) is sufficient to address problems of rural areas without adherence to principles of integrated rural development" (p. 145). They postulate that integrated rural development is essentially interventionist in nature, and incorporates the following elements:

- A multisectoral approach to development, with measures to promote sectors other than agriculture;
- Economic measures to be paralleled by initiatives in education, health, training and physical infrastructure investment;
- An attempt to target the most disadvantaged groups in rural areas;
- A requirement for people to become actively involved, not only in identifying needs and opportunities for development, but also in the implementation of projects;

- A demand for institutional reform expressed mainly as the devolution of powers from the national to regional and local levels of administration. (p. 147)

Martin and McKeown (1993) cite the example of Donegal, where investment in a digital network had not stimulated any notable economic development. They also note trends toward centralization of ICTs rather than distance-independent decentralization. "Even where decentralization occurs, the bulk of the jobs, control and decisionmaking, value added, and expertise remain at head office or in core area locations." They conclude: "Unless there is minimal infrastructure development in transport, education, health, and social and cultural facilities, it is unlikely that investments from [ICTs] alone will enable rural areas to cross the threshold from decline to growth" (p. 151).

Thus, in addition to availability of ICTs, other factors are necessary if they are to contribute to socioeconomic development:

- **Context:** It is important to understand the cultural, economic and political context. For example, if women are "invisible" in public or in commerce, they are not likely to learn to use ICTs without specific outreach activities. If there is no reliable transportation, producers may be forced to sell to a local middleman, even if they learn that prices are higher in the city. If there is no source of credit available, farmers may not be able to buy better seeds or pesticides, even if they learn from the Internet that these inputs would improve their crop yield. If there is no way to store harvested crops, farmers will not be able to wait to sell their crops even if they find out from futures markets that prices will be much higher in a few months.

- **Content:** Telephones do not require literacy or knowledge of a major language. "Infomediaries" can help to find and interpret useful content, but its true potential will be realized only if content becomes available in more languages, and more content relevant for developing regions is produced.

- **Capacity:** Use of ICTs requires the skills to use the equipment (such as typing for computers) and the ability to find useful information. Thus users require some training if they are to take advantage of access to ICTs. Alternatively, access sites such must be staffed by infomediaries who cannot only use the equipment but find relevant information, in order to assist new users, and these information brokers themselves will require training.

NOTES

1. *Creating a Development Dynamic: Final Report of the Digital Opportunity Initiative* (2001). New York: Markle Foundation.

2. See www.mssrf.org

3. See www.mtn.co.ug/smsinfo/index.htm

4. *Creating a Development Dynamic: Final Report of the Digital Opportunity Initiative* (2001). New York: Markle Foundation.

5. See www.ntclimited.com; see also www.peoplink.org

6. "*Defining Migration Policies in an Interdependent World.*" IOM Migration Policy Issues, No. 1, March 2003.

7. "Remittances: Sending Money Home." Inter-American Development Bank. See www.iadb.org/exr/remittances/

8. See www.gifttoethiopia.com. A "Holiday Special" includes a 30 kg sheep, a dozen red roses, and a bottle of scotch!

9. "Hispanic Banking: The Race is On." http://knowledge.wharton.upenn.edu/index.cfm?fa=viewArticle&id=1028

10. Robinson, Scott S. "Rethinking Telecenters: Knowledge Demands, Marginal Markets, Microbanks, and Remittance Flows." See www.isoc.org/oti/articles/0401/robinson.html

11. See, for example, Saunders et al. (1994), Hudson (1984), ITU (1986).

12. See Pierce and Jequier (1983) and ITU (1986).

13. "Healthnet." www.healthnet.org/healthnet.php

Chapter 3

Applications of ICTs
for Education and Training

We need information—masses of it. Without it, our culture will die.
—Inuit leader, Nunavut, Canadian Arctic

Perhaps the most extensive experience in using ICTs for development is in the field of education. The first applications of mass media to extend the range of education used radio in the 1950s and 1960s for nonformal education such as basic literacy for adults, and for formal education, either by teaching children directly in classrooms where teachers were unqualified, or by instructing the teachers themselves, so that they could better teach the students. In the 1970s, educators also began to use television in countries where it was available, such as American Samoa, India, and Côte d'Ivoire (Schramm, 1977). Today, radio is still used extensively for distance education in many developing countries. Televised courses may enable adults to study at home, whereas Internet-based courses offer the possibility for students to interact with each other and the instructor.

The advent of satellite technology offered an opportunity to distribute educational programs over an entire country or region. Satellites also provided a reliable means of interactive communication to link locations that previously had only high frequency (HF) radio or no telecommunications services at all. The potential of satellites and experiences in North America with experimental satellites (ATS-1 and ATS-6 in the United States, and CTS in the United States and Canada), led distance education specialists to propose projects that would use satellites for education in developing countries.[1] In the 1980s, two major initiatives using satellites for development were the USAID rural satellite program and Intelsat's Project SHARE.[2] In the 1990s, China used its domestic satellite system to deliver instruction to employees at their workplace, and in the United States, high tech workers could take graduate level engineering courses while students in rural high schools could study foreign languages or advanced placement courses via satellite.

However, the major change has been in the growth of interactive (as opposed to one-way broadcast) applications, ranging from real-time audio or video instruction and student interaction to Internet-based instruction and conferencing. Using the Internet, students can participate whenever they have time available, a significant advantage for instruction crossing time zones and for working students. E-mail and web-based instruction requires no more bandwidth than a dial-up telephone line, although more bandwidth is important if content includes complex graphics or video, or if many users share the channel, such as at a school or community telecenter. Yet even compressed video adequate for some instructional purposes may require as few as two digital telephone lines.

Applications of ICTs in education may be particularly beneficial for rural areas of developing countries, which typically have low literacy rates and a shortage of teachers and facilities. Similar, although less severe, problems face many rural parts of industrialized countries. As a U.S. Congressman stated at hearings on telecommunications in education, "The need for change in our education system is evident: massive illiteracy, low [test] scores and an acute shortage of teachers in rural areas."[3] Another educational goal is to reach the adult learner, both for basic education in the community and for training in the workplace. Here, a source of innovation is the private sector, which has made significant investments in ICTs for training its own employees to improve worker skills and reduce turnover. This chapter examines ICT applications and strategies that are intended to address educational challenges in rural areas of industrialized countries as well as in developing regions.

GOALS OF USING ICTs IN EDUCATION

As with applications in other sectors, ICTs may contribute to educational efficiency, effectiveness, equity, and access. For example, many uses of ICTs are designed to improving effectiveness or quality of education by providing better instruction or additional materials or interaction, to improve the learning experience for students with few classroom resources or poorly trained teachers, or for correspondence students studying on their own. In many cases, these ICT applications are also designed to improve equity by providing better learning opportunities for students in rural or impoverished areas. Another major goal is to extend access, by using ICTs to reach students who would not otherwise be able to study because courses are not offered locally. Of course, educational ICT projects may have multiple objectives, such as improving instruction and making curricula more relevant or extending reach and reducing cost per student.

The following are the most common goals of applications of ICTs in education:

- To improve the *quality* of instruction, for example by adding supplemental materials, research resources, opportunities to practice skills, interact with instructors, etc.;
- To extend the *reach* of education to people who have been excluded because of distance from educational institutions or disadvantage, for example, to reach isolated students, illiterates, people in remote jobsites or prisons, etc.;
- To *provide instruction* where no qualified teachers are available (for example, to teach foreign languages or advanced science or mathematics courses).

Often these goals are combined with the need to save *time* and/or *money*: for example, to enable employees to study at their workplace instead of traveling to classes or taking leaves of absence to attend distant courses; to enable a specialized instructor to teach students in several locations.

IMPROVING CLASSROOM INSTRUCTION

To improve the quality of formal education, the focus may be on the teachers or the students (or both). For example, a major problem may be that teachers do not have sufficient training to teach certain components of the curriculum. One approach is to use ICTs to teach students directly, with the local teacher serving as an aide or tutor. Radio has been used to teach basic mathematics to school children in Nicaragua, incorporating exercises using locally available materials such as sticks and bottle caps. Televised courses for schools were produced in American Samoa and Côte d'Ivoire. Alternatively, ICTs could be part of a strategy to improve the teachers' knowledge and skills by offering courses for teachers using correspondence materials, radio, television, and possibly the Internet, often coupled with incentives for completing credentials and earning pay raises. Instructional materials for teachers are generally less expensive to develop and produce than materials for young students with shorter attention spans. For example, Namibia chose English as its national language, but many of the teachers were not fluent enough to teach English in the classroom. Rather than developing distance education materials for the students directly, a teacher education project used a combination of radio broadcasts and print materials to improve teachers' fluency in English.[4] The project also produced English language radio programs to help with English comprehension. In some countries, sharing teachers who have the required expertise may be an option (see below).

ACCESS TO CLASSROOM MATERIALS

A common problem in developing regions is that there are not sufficient books for the students, or that the prescribed materials are dated. One solution would be to provide copies of textbooks and additional books on required subjects through a library in the school or the community. The Nakaseke community telecenter in Uganda used this approach to provide study materials for local secondary students, with books provided by UNESCO and the British Council. However, most telecenters or small community libraries would not have sufficient funds to purchase even single copies of required texts. An alternative solution would be to provide access to up-to-date materials for teachers to use in the classroom from online sources. For example, a teacher in Timbuktu (in central Mali) commented that the only African maps available in the school were from colonial times. That teacher could now use the telecenter's facilities to download and print an up-to-date map of Africa. Similarly, teachers could identify other needs and use the Internet to find current information and teaching materials. In fact, a teacher who managed a computer network in Zimbabwe noted that one of his major tasks was to compile a list of Web sites with relevant instructional resources, and to download and print out materials that teachers requested.[5]

Mass media such as films, videos, and television and radio programs expose students to history, science topics, and cultures beyond the classroom. A South African initiative is putting television sets in classrooms to enable teachers to use specially prepared programs for lessons on topics such as health and social skills. WorldSpace, a satellite system that transmits a wide range of radio programs directly to small receivers, plans to transmit educational radio programs via satellite for use in African classrooms and other settings such as clinics and community centers.[6] In the United States, an innovative project called LearnAlaska distributed educational programs to Alaskan village schools by satellite. The state government could not afford to produce televised materials for all subjects; instead, it asked teachers to list priorities for supplementary content, and then obtained educational rights to these programs and transmitted them by satellite at night when a channel was available. The teachers set their VCRs to record the programs, and then played them when appropriate in the classroom (see Hudson & Pittman, 1999).

The Internet can also enrich the classroom experience and provide learning opportunities not locally available. The WorldSpace satellite system can also be used to download information from Web sites, providing learning resources otherwise unavailable in African schools. The Internet and CD-ROM-based materials can even substitute for science labs. Inuit students in Arctic Canada use a virtual frog dissection kit available on the Internet (frogs are rare in the Arctic). Students in Uganda can collaborate on projects with students in other parts of the world using the Internet.[7]

ACCESS TO CONTINUING EDUCATION AND TRAINING

Another educational goal in developing regions is to upgrade the skills of working adults, such as extension agents, health workers, and employees of other government agencies. In industrialized countries, the working adults may be primarily private sector employees who cannot take the time away from their jobs to attend classes. This approach is sometimes referred to as nonformal education, that is, outside the formal educational system, although in some cases students may work to complete credentials from formal educational institutions. In the past, short courses that typically require participants to travel to major cities or take time from work, or correspondence courses that workers were expected to complete on their own time, were the major means of instruction. Today, such students may attend courses delivered through telecommunications facilities at their workplaces or local training centers, or may supplement traditional correspondence courses with online materials and assignments, and e-mail interaction with other students.

A variety of ICT-based strategies enhance correspondence courses. The goal is not only to enrich content, but to increase course completion rates through interaction with other students and feedback from instructors or tutorials. In many countries, print-based correspondence courses are offered for adult and distant students, who study on their own and mail in assignments and tests. Drop-out rates for such courses are often very high, as it may take a long time to get feedback on assignments by mail, and students who get behind or stuck on a particular topic tend to quit. Timely feedback and interaction with other students has been found to reduce dropout rates. For example, the University of the West Indies and the University of the South Pacific offer regular audio tutorials with instructors for correspondence students from their extension centers using audio conferencing and interaction through e-mail and group activities.

In Northern Canada, Keewatinook Okimakanak (K-Net) provides audio conferencing and e-mail for students in remote Ojibway and Cree villages of Northwestern Ontario who are taking correspondence courses to receive their high school diplomas. There are no high schools in the villages; teenagers must leave to attend high school in distant towns or cities. Many of the correspondence students are young people who dropped out of these high schools or whose parents did not want to send them so far from their families. In Australia, the School of the Air supplemented the correspondence courses sent out to children in the Outback, using two-way radios to keep in touch with students on remote farms or sheep stations. In addition to offering some instruction and enrichment, the radio provided a link with other children and "real teachers." The School of the Air now uses e-mail and on-line conferencing radio to reach children studying by correspondence on isolated homesteads.

In Peru, the Rural Communication Services Project (RCSP) was designed to use satellite communications via Intelsat to provide basic telephone service and teleconferencing to support development activities in an isolated region of Peru. Funded by USAID and managed by Entel Peru (Empresa Nacional de Telecomunicaciones del Perú), the RCSP provided public telephone service and audio conferencing facilities to seven communities in the Department of San Martin, a high jungle area east of the Andes. The conferencing activities were developed in cooperation with Peruvian agriculture, health and education ministries, and incorporated a wide variety of administrative, training, diffusion and promotion strategies. More than 650 audio teleconferences were carried out during the project. Entel itself became a major user of teleconferencing for training purposes. At the end of the two year period, Entel transferred responsibility for teleconferencing development to its commercial sector, with plans to promote teleconferencing among government agencies and private businesses (Mayo, Heald, & Klees, 1992).

JOB TRAINING FOR YOUTH

Another target audience for training in developing countries is teenagers and young adults. High youth unemployment is widespread; the problem may be lack of job opportunities for school leavers, or conversely, lack of appropriate preparation for available jobs. For example, the school curriculum may not provide skills in ICTs that are demanded by employers. The solution could be to introduce specific ICT-based courses into the curriculum if there are enough computers available and teachers with training to run such courses. A more appropriate approach may be to provide access to ICT training at school or in the community. Many initiatives that provide computers and Internet access to schools are intended to develop students' ICT skills. Community telecenters may also offer such training. For example, the primary activity of the Nabweru telecenter near Kampala, Uganda, is training in computer use; most of the participants are high school students and school leavers, and more than half are girls. In South Africa, some community computer centers are intended primarily to teach ICT skills to students and school leavers.

Some high tech companies support facilities and instruction in computer skills for young people in disadvantaged and developing regions. Cisco's Networking Academy Program evolved from a high school-based networking curriculum to a global e-learning program that provides web-based content as well as instructor training and hands-on labs, and prepares students for industry certification. Academies are based in high schools, technical schools, colleges and universities and community centers. Cisco provides facilities for these locations and uses a "train the trainer" approach at international and regional levels to prepare local academy instructors.[8] Other high tech companies such as IBM, Hewlett Packard, and Microsoft provide facilities and training in disadvantaged and developing regions. A key to training and certification is access to web-based materials so that motivated students can learn on their own. A young network manager at an adult computer literacy center in Soweto, South Africa, was introduced by the center director as having just completed his Microsoft Windows NT certification. Where did he go to take this course, he was asked. "I did it on the Internet" he replied with a grin.[9]

COMBATING ILLITERACY AND OTHER COMMUNITY PROBLEMS

A barrier to development may be low levels of literacy among children because there are not enough schools or teachers, or among adults who have not had a chance to go to school because of poverty, migration, or war. Again, two strategies can be used: teaching teachers or people who can become teachers, or teaching the target group directly. *Sesame Street* is an example of a television program designed to teach the basics of literacy and numeracy directly to young children; localized versions are produced in Mexico, Egypt, China, and South Africa, as well as several other countries. An alternative approach is to use ICTs to train more instructors who in turn can teach these target groups. Literate community residents without formal teacher training may learn how to present lessons to different target groups; they may also be provided with instructional materials in the form of audio cassettes, CDs, videos, or printouts of charts and worksheets to use in local literacy classes.

ICTs may also be included in strategies to address problems such as community health, which may involve high infant mortality, low birth weight, prevalence of disease due to poor sanitation, and prevalence of socially transmitted diseases. Radio and television are widely used for community education; soap operas (*telenovelas* in Latin America) are a popular format in many countries; *Sesame Street*'s South African version, *Takalani Sesame,* tackles fear and ignorance of HIV/AIDS; *Alam Simsim* in Egypt aims to build female self-esteem and encourage girls to attend and stay in school.[10] Canadian Inuit have established the Inuit Broadcasting Corporation which transmits TV programs via satellite across the North, including programs using puppets to teach Inuit children in their own language about health, hygiene, and northern living; and programs for teens and adults on Inuit culture and skills for survival in a changing world. The indigenous Imparja television network based in Alice Springs, Australia, also uses puppets based on bush animals in programs designed for aboriginal children in the Outback.

COMPUTERS FOR SCHOOLS

In recent years, many organizations have endeavored to place computers in schools, and to connect these computers to the Internet. Their implicit goals (rarely stated) are to improve the quality of instruction and to provide students with computer skills that they may need for future study and employment. School computer projects may range from schools equipped with one or two computers that may not have any communication, to a few computers of which at least one is connected to a modem for e-mail, to computer labs in schools or community centers with several networked computers and full Internet access. A variety of expectations may drive school networking initiatives, for example:

- The Ministry of Education may want to introduce ICT skills in the curriculum to prepare its citizens for the "information economy."
- Graduates with ICT skills may be more likely to get jobs than those without ICT skills.
- Resources available through ICTs may help students to understand difficult concepts in the curriculum.
- Information available through ICTs may help students to extend their knowledge and apply what they learn in the classroom.
- Interaction with distant students may increase understanding of other cultures and teach students to work in teams.
- Teachers may prepare more interesting and up-to-date lessons using information from ICTs.
- Integration of ICTs in the curriculum may foster a more exploratory and problem-solving approach to learning.

For example, a school networking project could be aimed at high school age students, to provide them with skills in using ICTs, to enrich the existing curriculum by providing access to resources on CD ROMs and the Web, and to enable them to collaborate on research projects with students in other schools or countries. Many factors will influence the extent to which these expectations are met, including extent of student access to the computers, attitudes of teachers and administrators, requirements of the existing curriculum, demands on teacher and student time, funds for connectivity and other operating costs, and so forth. These issues are examined in more detail in chapter 9.

Schoolnet Africa is an organization of school networking projects in many African countries, including Angola, Benin, Botswana, Egypt, the Gambia, Lesotho, Morocco, Mozambique, Namibia, Nigeria, South Africa, Uganda, Zambia, and Zimbabwe. The goal of Schoolnet Africa and other similar projects is not simply installing and networking computers, but improving education. As Isaacs points out:

- ICT usage in schools does not mean discarding the use of chalk and talk and textbooks;
- ICTs are not a universal panacea;
- It is not so much about the technologies as it is about education and how ICTs support educational outcomes.[11]

A major partner in school networking activities in developing regions is World Links with projects in Africa, Asia, Eastern Europe, Latin America, and the Middle East. World Links is a global learning network founded by the World Bank (and now an independent organization) that has implemented school-based learning centers in more than 1,000 communities. World Links connects teachers and students around the world for collaborative projects and integration of technology into learning. It also provides training materials and workshops designed to help teachers and students learn to use ICTs, and technology-enhanced curricula delivered on CD-ROM or via the Internet.[12]

DISTANCE EDUCATION

One of the most common educational applications of ICTs in developing regions is for distance education (i.e., reaching learners who are not attending classes in schools or on university campuses). ICTs in distance education may serve any or all of the goals listed above, such as to reach students outside the regular educational system; to save time and money for students, employers and/or educational institutions; to enhance the quality of correspondence courses; and to reduce dropout rates. There are several approaches to using ICTs in distance education:

• **The virtual classroom:** The most recent application of information technology for education at a distance is the virtual classroom—students may take courses anywhere they have an Internet connection, not only on a campus but at work or at home. Providers aggregate demand for education by reaching beyond the classroom and workplace to individuals interested in expanding their knowledge and/or obtaining educational credentials. Many educational institutions now offer courses and degrees online, although there is still controversy over whether the pedagogy and learning experience are equivalent to a college classroom. While people in developing regions are unlikely to be able to access the Internet from home, universities and professional organizations are delivering courses designed for students in developing regions who could access them at their workplace or through local schools or telecenters. Correspondence courses such as those run by UNISA (the University of South Africa) now include an electronic component for submitting assignments and interacting with other students. The African Virtual University, based in Nairobi, offers degree and certificate courses in English and French by partnering with traditional universities that provide the content. Courses include stand-alone electronic materials on CDs and DVDs, plus e-mail and web-based interaction. It has also established more than 34 learning centers in 19 African countries.[13] Indonesia's Open University is working with community Internet access sites known as *Warnets* so that students can participate online.

Broadcasters may also transmit televised courses for adult learners who study in the workplace (e.g., in China) or at home (more commonly in most countries). The Open University model typically includes hard copy correspondence materials plus some televised lectures. In the United Kingdom, face-to-face interaction with other students and tutors has been found to boost morale and reduce dropouts. Now many open university style programs include online interaction among students and with faculty. For example, In Canada, British Columbia's Open University offers more than 500 college and university level courses to more than 16,000 students, delivered through a variety of technologies including print, television, teleconferencing, and computer conferencing.[14] The University of Maryland and the University of Phoenix are among U.S. institutions offering online degree programs. Companies may offer their own ICT-based continuing education for employees. As noted earlier, Hewlett Packard, Cisco, Dell, and other multinational high tech companies provide ICT-based training for employees worldwide at a fraction of the cost of traditional classes.

• **The curriculum-sharing model:** This approach involves networking rural schools that may not have enough students to justify hiring specialized teachers in subjects such as mathematics, art, music, or foreign languages, or sufficient budget to do so. For example, in several rural areas in North America, schools are linked by microwave or optical fiber to share teachers and classes. Students at all the schools may take mathematics from a teacher at one school, and Spanish from a teacher at a different school. Instruction is typically by interactive video, so that the teacher may also see the students, although feedback via audio and text may also be used. However, organizational problems may be more significant than technical challenges. Schools may have to adjust class schedules to accommodate shared courses, and some may fear losing teaching positions.

• **The outside expert model:** This approach involves identifying course content that is not available in many rural schools, developing specialized instructional programming at a central source, and delivering the programs to the schools. These projects are typically regional or national in scope; many use satellites to transmit the courses to the schools and phone lines for interaction with students. For example, the TI-IN network was established in Texas to provide instruction via satellite to rural Texas schools that could not provide required high school courses in advanced mathematics, science and foreign languages. TI-IN then became part of Starnet, a related program to provide instruction via telecommunications to rural schools; Starnet was represented in more than 30 states, and used a variety of instructional technologies including live interactive satellite broadcasts, CD-ROM and web-based materials for instruction, course enrichment, and professional development for teachers.[15] STEP Star is another satellite-based educational network, that began as a multistate cooperative of school districts receiving educational programming, mostly in the western United States. It now offers interactive courses via satellite nationwide, including foreign languages, sciences, adult high school equivalency courses, curriculum enrichment, and professional development courses. Students who view the programs on DVDs rather than live may use the toll-free homework hotline to converse with instructional staff both before and after class hours.[16]

• **The consortium model:** In this approach, which has been used in higher education, several universities join together to deliver a complete curriculum to remote students at home or workplace. In the United States, the National Technological University (NTU) was a consortium of engineering schools at U.S. universities that delivered graduate technical courses via satellite to engineers at their workplaces throughout the country. Courses were also distributed via the Internet, CD-ROMs and videocassettes, particularly to overseas locations. Typically these courses were already being offered on a university campus and to the surrounding area via local television networks, using ITFS (fixed service microwave), so that there was little additional production cost or demand on faculty. Unlike the British Open University, there was no central campus or production facility.

NTU's participating universities received tuition revenue, but the degree was granted by NTU.[17] It appears that a version of this model has been adopted by the African Virtual University (see above).

• **The educational broker:** The broker serves as an intermediary to find existing content that may be of interest to educational institutions, or to professional associations whose members must fulfill continuing education requirements to retain their certification, such as physicians, dentists, and lawyers. The broker packages the content and leases the communications capacity (such as a block of time on a satellite). Participants may view the content at their workplace or at central locations such as hotels or conference centers that are equipped to receive the programs. In North America, the National University Teleconferencing Network (NUTN) offers a wide range of adult and continuing education programs from many sources.[18] Several other satellite conference brokers provide similar services. Typically, the receiving site pays a fee, and may then market the program in its region. Many professional organizations such as the American Medical Association, American Bar Association, and U.S. Chamber of Commerce use such facilities to offer continuing education to their members.

ICTs IN EDUCATION: FINDINGS AND ISSUES

Most research on educational applications of ICTs focuses on comparing effectiveness with face-to-face instruction. Students studying at a distance tend to score about the same as counterparts in face-to-face classes. In many cases, they perform better than students in traditional classes. Witherspoon, Johnstone, and Wasem (1993) suggest two reasons for this result: Offsite students are typically older and more motivated than students on campus, and classes taught using technology are frequently designed more systematically to create a successful learning experience. However, Schramm's (1977) conclusion in the 1970s that motivated students can learn from anything ranging from chalkboards to television appears to hold also for more recent applications such as videoconferencing and Internet-based instruction.

Although research on the actual costs and benefits of educational technology is much more limited, some data is available from corporate users of training networks. For example, Hewlett Packard's Information Technology Network (ITE-Net) provided interactive voice and data communication with employees in more than 100 classrooms worldwide. HP estimated that its Distance Learning System delivered training at one-half the cost of traditional classes (Portway, 1993). A study by Cronin and colleagues addressed the question: "What cost savings associated with the education sector's usage of telecommunications can be expected?" Using a translog cost model,

Cronin, Gold, Mace, and Sigalos (1994) calculated cost savings in the U.S. educational services sector due to advances in telecommunications production and education's consumption of telecommunications for each year from 1963 to 1991. Cumulative cost savings totaled $76.7 billion in 1991 dollars for U.S. applications. However, the authors suggested that the full potential of telecommunications as a substitute for more expensive inputs and processes had not been realized, as education's real usage of telecommunications had increased by only 1.8 percent per year, less than half the annual national rate of 3 percent. They concluded: "Through distance learning programs, telecommunications may efficiently promote a more equitable distribution of educational and informational resources among the relatively resource poor" (p. 74).

Another approach is to examine the impact of ICT use on student drop-out rates. In some correspondence programs for high school and university courses in developing regions, drop-out rates may exceed 75 percent because students get behind, lose motivation, or have other pressing priorities. High drop-out rates represent wasted educational resources, as well as developmental setbacks, since many correspondence students in developing countries are trying to obtain credentials for teaching, health care positions, or other employment. These data rarely show up in evaluations, which typically compare the grades of students who do complete the courses with grades of students in typical classrooms. However, use of telecommunications for distance learning may reduce student drop-out rates.

The University of the South Pacific operates a satellite-based audio conferencing network linking its main campus in Suva, Fiji, with its agricultural college in Samoa and with extension centers in ten other Pacific island nations. The system is used for administration of extension services activities and courses, tutorials for students taking correspondence courses, and outreach services to bring the resources of the University to the people of the region—through consultation, in-service training, seminars by UN and other development agencies, etc. The benefits of this experimental network have been significant. The savings in travel time and costs resulting from having meetings over the network rather than bringing a representative from each location to Fiji, have been at least ten times the cost of using the network. Drop-out rates of correspondence students in courses with effective satellite tutorials have also been reduced (Hudson, 1990). In northern Canada, Keewaytinook Internet High School uses ICTs plus local tutors to help increase completion rates of native students in remote communities who are studying by correspondence to obtain their high school equivalency diploma.[19]

Similarly, access to distance education may contribute to reductions in staff turnover in remote locations. Anecdotal evidence from remote northern settlements in Canada and Alaska and from health care professionals in rural Texas indicated that access to education to improve their qualifications and upgrade their skills was an important factor in determining how long professional staff would remain in isolated areas. Costs associated with high turnover include not only recruitment and relocation, but lost productivity while positions are unfilled and new staff adapt to local conditions.

Evaluation of Internet Access for Schools

The general premise of school networking projects is that access to computers that are networked for e-mail and Internet access can be beneficial for education. However, this premise is based on many assumptions about access, skills, and applications. Faulty assumptions can result in few significant benefits from the project. For example, if the computers break down frequently or the telephone line is unreliable, the schools will not be able to put these tools to use. If there are so few computers that students can use them for only an hour a week, or if the computers are locked up when the students are free to use them, there will be little benefit. If the schools cannot afford installation or access charges for telecommunications, the benefits of sharing information by e-mail and finding information on the Internet will not be achieved. Even if the facilities work reasonably well and are relatively accessible, if the teachers are not able to incorporate their use in the classroom and the curriculum, much of the project's potential benefit may not be realized (although motivated students who learn to use the computers may benefit to some extent). The author has witnessed all of these problems in the field!

World Links has conducted some evaluation of its projects that provide Internet access and connect schools in different countries, focusing on ICT use by girls, ICT skills, and career interests among participating students. Comparative evaluation, including students and teachers in schools with and without World Links projects, found that World Links students' general knowledge about computers, the Internet, and software increased, and that they gained in knowledge about other cultures and communication skills. Boys' and girls' uses of computers and Internet access differed somewhat, with girls focusing more on academic research, collaborative work and e-mail, reproductive health and sexuality, and subjects considered "taboo" in their culture. Both boys and girls gained knowledge about other cultures and improved their attitudes toward school. Girls also gained in reasoning and self-esteem. In Uganda, female university students who had participated in World Links school projects were more computer literate, more likely to use ICTs in their studies, and more likely to be studying for an ICT-related degree and planning a career in an ICT-related field.[20]

In the United States, evaluation of the Universal Service Fund's E-rate program that provides discounted Internet access to schools and libraries has focused primarily on access and usage rather than the effects of Internet usage to date. Findings by Puma et al. (2002) show that "Students in poorer E-Rate districts and schools are . . . more likely . . . to face a variety of conditions that may limit their use of technology for instruction, including inadequate teacher skills, limitations of school buildings . . . and speed and reliability of Internet connections" (p. viii). These problems exist in both rural and inner city schools, and in large schools and districts where resources in terms of equipment, support staff, and trained teachers may be stretched very thin. These preliminary assessments indicate that bridging the educational digital divide requires much more than equipment in both developing countries and disadvantaged regions of industrialized countries.

Sustainability

Sustainability, or the ability to continue an activity after the experimental or pilot phase, remains a daunting problem for educational ICT initiatives. Funds for spare parts, maintenance, and supplies such as toner cartridges may be beyond the budget of many schools, even if equipment was donated. There may be no funds to pay for technical staff or to contract for technical support. Internet access fees are likely to be prohibitively expensive if not subsidized indefinitely. Not only are schools short of funds for basics such as textbooks, writing materials, and electricity, but the cost of maintaining even donated computers and of Internet connections may be unbearably high. The cost of $22 per student per year estimated by World Links may be low, but beyond the budget of most schools. World Links states that the average recurring costs for Internet access and networked computers of about $250 per month per school are typically raised by parents and others in the community rather than coming from the Ministry of Education. Some schools also raise money by providing public access at a fee. World Links states that 98 percent of the schools it has supported are still online.[21]

Organizational barriers may also significantly inhibit sustainability. Where resources are shared, institutional rivalries and other constraints may be far more challenging than providing training or technical support. A telephone company that built a fiber network connecting schools for curriculum sharing found that schools preferred to lobby for more teaching positions rather than sharing specialized teachers. Where students are scattered across many timezones, live lectures with realtime interaction with the instructor (the "candid classroom" model) may be too inconvenient for many participants. If courses offered by one institution are accessible to students from other institutions, how are the tuition revenues allocated? Which institution gives the degree or credential? Such problems have stymied several proposed collaborative degree schemes.[22] None of these barriers is insurmountable if there is sufficient will, or incentive, to make the ICT project work. Yet overcoming them is critical if effective ICT applications are to move beyond demonstrations and pilot projects to being integrated in the instructional process.

Rankin Inlet: An Arctic Window

In Rankin Inlet, a Canadian Inuit (Eskimo) settlement on Hudson Bay "800 miles from everything," more than one in five residents has an e-mail account. At night a school classroom becomes a community access center called Igalaaq, meaning "window" in Inuktitut.[23]

The Internet adds to the curriculum as students can exchange information with children in the Australian Outback and Hawaii, and study biology using the Virtual Frog Dissection kit. But the links have the potential to strengthen indigenous culture and institutions as well. An elder teaching students to sew clothing from seal and caribou skins uses a digital camera to record their work for posting to their website.

Canadian Inuit have used telecommunications in the form of telephone, fax, and videoconferencing to lobby for land claims and greater political autonomy. The eastern part of the Northwest Territories became a separate Inuit-run territory called Nunavut ("our land") in 1999. Northern leaders see the Internet as another powerful communications tool to unite the isolated Inuit communities. Indigenous peoples who call themselves the "fourth world" are also using the Internet to link aboriginal organizations throughout the world to share concerns and organize to press for human rights and environmental issues.

Northerners also see the Internet's potential for economic development. Sakku Investments, the business development arm of the local Inuit association, donates equipment and access time, viewing the Internet as the electronic road system for their business development. "As far as I'm concerned," says Sakku's CEO about Igalaaq, "it's a driver's ed school."[24]

Linking Remote University Extension Centers: USP and UWI

The University of the South Pacific uses a satellite-based network called USPNet to provide tutorials to correspondence students scattered in 12 island nations of the South Pacific (Cook Islands, Fiji, Kiribati, Marshall Islands, Nauru, Niue, Samoa, Solomon Islands, Tokelau, Tonga, Tuvalu, and Vanuatu). In 2001, some 62 percent of the certificate and diploma students and 26.5 percent of degree students were external, typically studying by correspondence, with help from local tutors and the resources of USPNet.[25]

The USP Communications Network was established in 1974 to provide a communications system to help bridge the vast distances between the main campus in Suva, the other USP campuses in Samoa and Vanuatu, and local extension centers. The networks evolved from initial use of HF radio and the NASA Peacesat satellite (with a single VHF channel and a footprint reaching from the South Pacific to Alaska) to a dedicated VSAT network owned and operated by the University. USPNet enables distance students to participate in audio tutorials conducted from any campus, communicate by e-mail with a lecturer/tutor or another student, access the World Wide Web, watch a live video broadcast of a lecture from any of the three campuses (in Fiji, Samoa, and Vanuatu) and take part in interactive video conferences and tutoring) with the main campus in Fiji. Also the network's communication facilities are important in saving time and travel for the staff of the correspondence program and the extension centers scattered across several time zones (and both sides of the International Date Line) and thousands of miles of ocean. General university administration has also become more efficient with e-mail communication via USPNet to all USP locations.

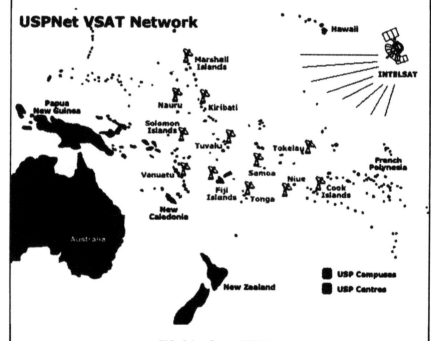

FIG. 3.1. *Source.* USPNet.

In the Caribbean, the University of the West Indies established an audio teleconferencing network called UWIDITE (UWI Distance Teaching Experiment—later Enterprise) to link its campuses in Jamaica, Barbados, and Trinidad with extension centers throughout the Commonwealth Caribbean (Block, 1985). This initiative has evolved into UWIDEC (for UWI Distance Education Center) headquartered at the Trinidad campus, and with facilities at the main Jamaica campus and a smaller unit at the Barbados campus. The audioconferencing network links the three campuses with university centers in 15 other Caribbean nations. The network may be used for lectures, tutorials, and meetings, enabling students to complete certificates and to participate in outreach programs, as well as to listen to lectures in regular UWI courses in preparation for taking exams to complete the first part of their degrees from their home countries. UWIDEC now provides additional materials and support for distance education in the region.

NOTES

1. For an analysis of the North American experience with experimental satellites in the 1970s, see Hudson (1990).
2. See Tietjen (1989) and Hudson (1990).
3. David Markey, quoted in U.S. Congress (1993).
4. Commonwealth of Learning. "Using Community Radio for Non-formal Education." www.col.org/knowledge/ks_radio.html
5. Personal interviews by the author.
6. See www.worldspace.com
7. See www.world-links.org
8. See http://cisco.netacad.net/public/academy/About.html
9. Personal interview by the author.
10. Hawthorne, Peter. "Positively Sesame Street." Time Europe, Sept. 22, 2002. http://www.time.com/time/europe/magazine/article/0,13005,901020930-353521,00.html
11. Isaacs, Shafika, "ICTs in African Schools: The Experience so far." See www.schoolnetafrica.net
12. www.world-links.org
13. See www.avu.org
14. In April 2005, BCOU became part of Thompson Rivers University. See www.bcou.ca
15. Starnet is now part of the United Star Distance Learning Consortium (USDLC). See www.usdlc.org
16. STEP originally stood for Satellite Telecommunications Educational Programming. http://stepstar.esd101.net
17. See www.ntu.edu
18. NUTN is now hosted by Old Dominion University. See www.dl.odu.edu
19. See http://kihs.knet.ca
20. Carlson, S. and McGhee, R. "World Links . . . Opening a World of Learning." June 19, 2002. http://world-links.org/english/assets/NECC2002.ppt

21. See endnote 20.

22. Such problems led to the restructuring of NTU, which became part of Walden University in 2005.

23. MacQueen, K. "Surfing the World from the Frozen North." Ottawa Citizen, February 23, 1997. www.schoolnet.ca/aboriginal/learningcircle/05/frozen-e.html

24. See endnote 23.

25. See USPNet, www.usp.ac.fj/its/sections/uspnet/index.html and University of the West Indies Distance Education Centre www.dec.uwi.edu/

Chapter 4

Telemedicine and Telehealth: Applications of ICTs for Health Care Delivery

Information is the fuel of medicine. Here we have none. Year by year we are falling behind.
—Doctor in Timbuctu, Mali, before a telecenter was established

THE CONTEXT

Developing regions typically face severe shortages of physicians, particularly in rural areas. In general, health care in developing countries must be dispensed by individuals with less training and less backup than their counterparts in the industrialized countries. Health workers may have only minimal training, or have few opportunities to upgrade their knowledge and skills. Specialists may be available only in the major cities. Facilities for treating patients may be inadequate in terms of staffing, equipment, and medications. The health sector in developing countries also confronts major administrative, quality control, and logistical problems.

In addition, mortality and morbidity rates are generally higher than in urban areas due to poor sanitation and other environmental conditions, and dangerous occupations. Although the problems they face are much less acute, rural areas of industrialized countries also face difficulties in recruiting and retaining health professionals, and typically have higher mortality and morbidity rates than urban areas. An Office of Technology Assessment (OTA) study cited three problems specific to residents of U.S. rural areas:

45

- **Health indicators:** a disproportionate number of rural people suffer from chronic illnesses; the infant mortality rate is higher than in urban areas; the number of deaths from injuries is dramatically higher;
- **Distance from care:** lack of transportation and few local providers make it difficult to reach health care facilities;
- **Poverty:** poverty is higher in rural areas than in the nation as a whole (Witherspoon, Johnson, & Wasem, 1993, p. 1).

Structure of the Health Sector

Primary health care is provided to an increasing degree in developing countries by paraprofessionals, who are often selected from the local community, or may be sent to rural clinics from urban centers. They may be community health workers or health aides with a few weeks or months of training, medical assistants with about one year of training, or practical field nurses. The duties of primary health care workers vary widely by country, but generally focus on preventive, promotive, and simple curative care. This includes early diagnosis and treatment of common illnesses, maternal and child health care, midwifery, family planning, treatment of injuries, and referral of patients to higher level facilities, if available. Primary health care workers may also organize immunization and mass treatment programs; provide guidance and education on nutrition, family planning, and hygiene; monitor epidemics, water quality, and sanitation; and collect demographic and health information (Evans, Hall, & Warford, 1988).

The second level of health care varies according to location and the complexity of the health care system. However, it normally refers to services available at larger health centers and small district or regional hospitals. Such services may be dispensed by nurses, paramedics, or general practitioners. The third level of care is normally the most sophisticated technologically and has the greatest mix of health personnel and diversity of services. It usually includes medical specialists based in major hospitals in larger urban centers. In some countries, highly specialized hospitals treating specific health problems, such as leprosy or mental health, constitute in effect a fourth tier. Although we might think of health providers structured as a pyramid with a large number of paraprofessionals on the bottom and a few specialists at the top, some developing regions may have a different distribution. In Latin America, instead of resembling a pyramid, the structure of health manpower "resembles a sand clock with a slip neck of technicians and two receptacles representing a cadre of auxiliaries and a cadre of professionals" (Golladay, 1980, p. 10). This shortage of technicians in particular is common in many developing countries.

Information and communication technologies are obviously not a panacea for all health care problems. However, they can play an important role in improving access to care and quality of care. The most obvious example of quality or effectiveness is in emergencies, when quick communication to get help can save lives. However, more mundane applications may reap larger benefits—such as getting advice on treating a patient with an ailment that cannot be diagnosed locally, ordering medicines to be delivered before the clinic runs out, coordinating visits of itinerant health professionals so that village women do not miss getting PAP tests and children don't miss inoculations.

TELEMEDICINE AND TELEHEALTH

This chapter examines the various applications of telecommunications in support of health care, with particular relevance to rural and developing regions. Generically, these applications are referred to as *telemedicine*, although some researchers and practitioners prefer to use that term for consultative uses, and the term *telehealth* to refer to applications for medical education and administration. Initiatives using ICTs (information and communication technologies) to support health services include the following:

- **Emergencies:** to summon immediate medical assistance; to communicate with emergency vehicles and staff;
- **Consultation:** typically between primary health care providers and district level physicians, or between district physicians and specialists;
- **Remote diagnosis:** for example, transmission of medical data and images, interpretation of data by distant specialists;
- **Patient monitoring:** for example, transmission of patient data from home or rural clinic, often coupled with followup through local medical staff;
- **Training and Continuing Education:** of health care workers, paraprofessionals, physicians, etc.;
- **Public Health Education:** of target populations including expectant mothers, mothers of young children, groups susceptible to contagious diseases, etc.;
- **Administration:** ordering and delivery of medications and supplies; coordination of logistics such as field visits of medical staff; accessing and updating of patient medical records; transmission of billing data, etc.;
- **Data collection:** collection of public health information such as epidemiological data on outbreaks of diseases;

- **Research and information sharing:** such as access to medical databases
  and libraries and consultation with distant experts and peers.

Emergencies

Emergency communications offer the most dramatic evidence of the importance of communications in health care delivery because getting help quickly can save lives. Studies in India, Costa Rica, Egypt, and Papua New Guinea, all showed that about 5 percent of rural calls were for emergencies and medical reasons (Hudson, 1984). The indirect benefits of these calls in terms of saved lives and reduced suffering are highly significant. In natural disasters, such as earthquakes and floods, and for epidemics, two-way communication systems (by mobile telephone, two-way radio, or portable satellite links) are used to secure assistance and to coordinate the logistics for emergency relief.

Two-way radios have been used in many developing countries to coordinate disaster relief activities. Now portable satellite stations may be brought in on trucks or small planes to provide reliable emergency communications. Portable satellite terminals were used for relief activities following the 1980 Mount St. Helens eruptions in the United States (Goldschmidt, Forsythe, & Hudson, 1980) and the 1986 earthquake in Mexico City. In the South Pacific, the World Health Organization has used the experimental PEACESAT satellite network to summon medical teams during outbreaks of cholera and dengue fever, and to coordinate emergency assistance after typhoons and earthquakes (Hudson, Goldschmidt, Hardy, & Parker, 1979). Handheld satellite phones (using low earth orbiting satellites—see chapter 7) are now used in responding to natural disasters and other major emergencies, such as the tsunami in the Indian Ocean in December 2004. However, these satellite phones remain too expensive to be included as standard equipment in rural clinics. In urban and periurban areas, the proliferation of cell phones has made significant improvements in emergency communications. Mobile phones can also replace regular wireline telephone service when cables are damaged, such as after the 1989 San Francisco earthquake and the 1995 Kobe earthquake.

Emergency location beacons transmitting signals to a satellite enable rescuers to find ships lost at sea or downed aircraft. Portable Global Positioning System (GPS) units carried by emergency workers can pinpoint disaster locations to provide guidance for logistics support, such as medical evacuation and delivery of medication, tents, and food. Telecommunications systems can also help to prevent disasters. For example, following several severe cyclones that caused great loss of life in Bangladesh in the mid 1970s, the government implemented a cyclone early warning telephone system consisting of single telephone installations in several coastal areas previously without telecommunication access. Tsunami warning systems were being planned for the Indian Ocean following the December 2004 disaster.

Consultation

As noted above, developing countries now rely on paraprofessionals for delivery of basic health services, particularly in rural areas. These health workers receive basic training in treatment and prevention of common health problems, but need supervision and assistance in diagnosing and treating uncommon diseases and coping with serious health problems. Telecommunications links between village clinics and regional hospitals or health centers can be used for consultation and supervision. Where telephone service is not available, two-way radio networks are still used in many developing countries to support isolated paraprofessional health workers. Some birth attendants in Uganda can summon help from rural clinics using portable VHF radios. In Guyana, rural health workers called "medex" use a two-way radio network to communicate with headquarters in Georgetown to check on delivery of drugs and supplies, and to receive advice on major health problems. They may also request emergency evacuations and follow-up on patients referred to hospital.

Satellites can provide more reliable communications for remote and isolated regions. In Alaska, village health aides are in daily communication via satellite with physicians at regional hospitals. More than 100 Alaskan villages are equipped with earth stations that are used for the dedicated medical network, long-distance telephone service, and television reception. Alaskan projects are now adding personal computers and auxiliary equipment to enable health aides to transmit otoscope images, heart sounds, and other patient data as e-mail attachments to regional hospitals. Doctors at these hospitals, in turn, can transmit x-rays and other medical data to specialists at a regional medical center (Hudson & Pittman, 1999). See boxed text on pp. 59–60.

SatelLife of Cambridge, Massachusetts, operated a store-and-forward satellite system, using a low earth orbit (LEO) satellite, HealthSat-2. The satellite's unique polar orbit allowed ground stations to transmit and receive data from any point on earth daily. Stations close to the equator acquired the satellite four times a day; with each acquisition or "pass" lasting for about 13 minutes. SatelLife's Healthnet Knowledge network enables medical practitioners to seek advice on treatment of unusual cases from colleagues in other parts of Africa, and to download articles from medical libraries. Burn surgeons in Mozambique, Tanzania, and Uganda have used Healthnet to consult with one another on patient treatment and reconstructive surgery techniques. In response to a cholera epidemic in Zambia, the medical librarian at the university obtained literature from her "partner library" at the University of Florida, and then disseminated the information to all Healthnet users in the region. Healthnet is also experimenting with handheld computers or personal digital assistants (PDAs) for data collection and information for rural health workers in Uganda.[1]

Physicians in rural and developing regions may also seek real-time advice from distant medical schools. Memorial University of Newfoundland has provided assistance to isolated physicians in Newfoundland and Labrador via audio and computer links. Project Rainforest, a collaborative project of Yale University and the National Aeronautics and Space Administration (NASA), conducted a unique experiment in telemedicine, connecting the jungle of eastern Ecuador with the consulting resources of Yale Medical School. Laptop computers and telephone lines linked a mobile surgery program in Cuenca, high in the Andes, with an isolated hospital ten hours away in Sucua, Ecuador and with surgeons on the Yale campus. The Ecuadorian doctors in Cuenca examined their patients in Sucua and discussed their care with primary physicians using telephone lines. The team then transferred to Sucua where a laparoscopic procedure was performed with real-time consultation and monitoring from the Yale campus.[2]

"E-consultation" using the Internet can link practitioners to specialists around the world. ORBIS International, which provides eye care in developing regions, has started an e-consulting program connecting a partner ophthalmologist located anywhere in the world with an ORBIS telemedicine mentor (Telementor) for professional assistance. The system can be used for training using online cases and commentaries, or for patient management, with digital images and patient histories being sent from the ophthalmologists to the telementors for advice.[3]

Medical services in some countries use small planes to take physicians to field sites rather than bringing patients to central hospitals. This approach, known as the "Flying Doctor" began in Australia in 1928. People living on remote ranches or "stations" used two-way high frequency radios to contact regional Flying Doctor bases. Doctors flew in to provide assistance, often piloting the small bush planes themselves. Fifty years later the "mantle of safety," the area served by the Flying Doctors, covered more than 80 percent of Australia (Page, 1977). Today, the Flying Doctor Service continues, but high-frequency radios are being replaced with telephone service provided either terrestrially (typically via digital microwave links) or via satellite.

Similar health communication networks are found in other parts of the developing world. Flying doctor services in several east African countries (including Kenya, Tanzania, and Malawi) use two-way radio networks to link nurses at rural clinics with headquarters, and to coordinate the aircraft used to transport doctors to the clinics and to evacuate seriously ill patients.

Two-way radio is a vital support for this service, with radio telephone communications linking the headquarters with all field hospitals and clinics and with the airplanes themselves through installation of a common frequency in the planes' radios. Air evacuations arranged by two-way radio include a young woman in labor requiring a caesarian section, a newborn with a large tumor, and a farmer gored by a water buffalo.[4]

Training and Continuing Education

Telecommunications and information technologies can also be for training and continuing education of health workers. In Alaska, the printed health aide training manual is now also available on CD-ROM, including graphics and interactive training materials. Memorial University of Newfoundland's Telemedicine Program has developed a distance education program delivered to rural physicians using the combined technologies of CD-ROM and the Internet; video, audio, and still images would take 20 minutes to download if they were Internet-based. For this education program, a brief signal instantly triggers physicians' computers to play the CD-ROM they have already received. A similar approach may be appropriate for developing countries, where bandwidth is limited and costs of Internet access are often high.[5]

Interactive audio conferencing can also be used for continuing education of health workers. Research in Alaska found that health aides learned from participating in "doctor call" in which each aide in turn presented cases to physicians at regional hospitals over the experimental ATS-1 audio satellite system. This experience led to audio-conferencing being included in the commercial satellite network designed to serve all Alaskan village clinics (Hudson, 1990). The Peru Rural Communications Services Project sponsored by USAID connected health workers in the eastern jungle of San Martin to Lima via VHF radio and satellite links for consultation and continuing education (Hudson, 1990; Tietjen, 1989). In Guyana, training staff at the national hospital in Georgetown ran interactive refresher sessions and "grand rounds" presentations by physicians of medical cases over the high-frequency radio network (Hudson, 1984). Since all clinics shared the same frequency, participants could hear questions and comments from all the sites, in a format similar to Alaska's satellite-based "doctor call."

In addition to providing information, another benefit of telecommunications cited by rural health workers is reducing the feeling of isolation from colleagues. Researchers have hypothesized that reducing isolation can help to reduce personnel turnover. While causal data are difficult to obtain, it appears that communication is at least an important factor. For example, medex in Guyana reported that chatting over the radio in the evening helped to reduce their sense of isolation and boost morale (Fryer, Burns, & Hudson, 1985). Better communications is cited as one of several factors encouraging reversal of the medical braindrain in Navrongo, Ghana.[6] Memorial University's Telemedicine Project also states that one of the main recruitment issues for rural doctors is availability of continuing medical education.[7]

Public Health Education

Numerous projects have used communication technology in support of health education, much of which is targeted to women as expectant mothers, mothers of small children, and community residents. Successful campaigns have used a variety of media from posters to radio and television, depending on the message and the resources available. Recent radio campaigns have targeted HIV/AIDS and malaria prevention and family planning. Interactive communication use is less common, but may be involved for administrative support and for follow-up with health workers and project staff.[8]

Television programs for continuing education of health staff and for public health education (both to patients in hospitals and the general public) are increasingly available in the United States. Channels offering educational programs and entertainment for patients are received via satellite and distributed via internal cable systems to patients' rooms. Continuing education channels for health care professionals are also distributed nationwide via satellite (Hudson, 1990). Satellite networks such as Worldspace could be used to distribute public health radio programming in many languages throughout Africa.

Yet not all families in developing regions have access to broadcast services, and messages may be more effective if delivered in person and customized for local audiences. For example, community health workers may play videos to show mothers how to prevent and treat childhood diseases or to increase awareness of how HIV/AIDS can be transmitted. Other community initiatives such as improving sanitation and nutrition may also involve ICTs to enable community leaders to obtain information on practices adopted elsewhere or to demonstrate new techniques to community groups. However, research from campaigns in both industrialized and developing countries has shown that the use of ICTs to increase awareness must be coupled with strategies to change behaviors if campaigns are to be successful. In general, it has been found that mass media messages **plus** interpersonal communication are needed. For example, mothers can learn from mass media campaigns that it is important to rehydrate babies who have diarrhea. But learning directly from a community health worker how to prepare a specific mixture of clean water, salt, and sugar is likely to result in greater adoption of oral rehydration therapy (Meyer, Foote, & Smith, 1985). Similar strategies involving mass media plus interpersonal follow-up are now being used in AIDS prevention campaigns.

Administrative Support

Another important use of health communication systems is for administrative functions, including ordering drugs, checking on delayed shipments of supplies, transferring patient record information, coordinating staff travel, and evacuating patients. In Guyana, for example, the dedicated two-way radio telephone network was used most frequently for administration. Medex with two-way communication facilities received drugs more quickly and kept a more complete supply than medex without two-way radios. Also, administrative problems that took weeks or months to resolve by mail or personal visit were resolved within hours by radio telephone (Meyer et al., 1985). Physicians in Ethiopia use Healthnet to schedule consultations and referrals, making it unnecessary for ill patients to travel long distances with no guarantee of seeing a physician.[9] The African Medical and Relief Foundation (AMREF) based in Nairobi, uses two-way radios, mobile phones, satellite phones and e-mail to coordinate evacuations and schedule doctor visits and training sessions.[10]

Computerized record keeping systems are now also used by many agencies in several countries, including the Public Health Service in Alaska and in the southwestern United States to store and update patient data. A feature of the system is that patient records can be accessed and updated at different locations, which is particularly advantageous both for migratory populations who appear for treatment at various clinics and for keeping track of patients transferred within the system. Before visiting native villages, practitioners in Alaska can access medical records to compile lists of children who need vaccinations and women who need PAP smears. Computerized records can also be sorted and grouped according to any variable so that it is possible to monitor, for example, heart patients with histories of rheumatic fever, and to provide up-to-date lists for use by itinerant health workers of children requiring vaccinations or women requiring periodic checkups. The electronic availability of patient records, which requires very little bandwidth, could have a significant impact on efficiency of patient treatment, and reduce guessing and errors in treating patients with unknown medical histories.

Although diagnostic applications attract the most attention, administrative tasks such as computerizing medical records and claims processing (for billing to private insurance carriers or government insurance schemes) may yield the greatest cost savings in industrialized countries. In the United States, pharmacy chains now share a nationwide database so that customers can obtain prescriptions at any location. Independent pharmacies or health care providers may join together electronically to get better prices for their customers than they would be able to obtain individually. Online services make it possible for pharmacies, hospitals, physicians, and dentists to exchange data electronically with claims processors, reducing paperwork and time to process health insurance claims (Stewart, 1994a).

Data Collection, Research and Information Sharing

Communication systems can also be used to collect field data on outbreaks of infectious diseases and to issue medical alerts. In Mali, the Radio Administrative de Communication (RAC) network links clinics and hospitals to Bamako to collect epidemiological data. In The Gambia, health workers who once had to travel 700 kilometers per week to collect data for a clinical trial now use Healthnet to send this information via electronic mail. Health care workers in the Congo's Vanga Hospital use Healthnet to send regular dispatches to report on progress in treating trypanosomiasis to health organizations in the North. Malaria researchers at a remote site in northern Ghana used Healthnet to communicate daily with the London School of Hygiene and Tropical Medicine and the Tropical Disease Research Center in Geneva.[11]

ICTs also now provide opportunities for medical students, practitioners, and researchers to keep up with current research, track down the latest information on illnesses and injuries they must treat, and share information with the emerging global virtual health care community. They may consult with other physicians via e-mail or search online libraries for guidance. For example, a physician was able to avoid amputating the leg of a young athlete infected with necrotizing fasciitis (flesh-eating bacteria) after finding information on an alternative procedure in the MEDLINE online medical database (ITU, 1999). This latter function of information-sharing, now expanding in developing countries via Internet access, may be the most valuable of all medical applications. There have been examples of physicians in China as well as Africa and other developing regions successfully treating undiagnosed patients after seeking advice over the Internet.

Supervision and Quality Assurance

The paucity of rural health care professionals and the decentralization of services among many locations exacerbate the problems of maintaining consistent quality of service in all locations. Hence, the supervision of medical performance and the monitoring of the utilization of services offered become critically important. The most effective form of supervision is periodic visits by more senior personnel; however, in many countries difficult terrain or shortages of vehicles, fuel, and expert staff make supervision visits infrequent. Where available, the telecommunications system has been used to supplement personal contact by scheduling regular periods for telephone contact with each health worker to monitor or check on procedures, and on occasion, to prompt the health worker about specific functions and duties. Such frequent and interactive contacts with health workers may help to maintain the quality of service dispensed by individual workers and to reduce regional disparities in quality of service.

Another facet of the administration problem is that the management expertise necessary to organize and operate a complicated system of health services for rural and semiurban populations is in very short supply in developing countries (Evans, Hall, & Warford, 1988). With rapid two-way communication, the limited management expertise available can be more widely effective and proactive in managing available resources, rather than simply reacting to problems as they emerge.

Morale and Confidence

An important factor in the acceptance of primary health care by rural populations can be a knowledge that a doctor or nurse is "on call" by telephone and that they can back up the local health worker's advice to a patient, and hence boost the patient's confidence in the accuracy of advice or assistance being given. Rural health aides have reported that they feel more confident in their diagnoses and treatment of patients, and have found it easier to gain patient compliance when their advice was backed up by consultation with a physician (Hudson & Parker, 1973).

A second important factor in the success of a developing country health care system involves the morale of field workers dispensing primary health care. The ability to maintain daily contact with health workers in other villages and other sectors of the system reduces feelings of isolation and boosts the sense of security and the morale of health workers. For example, in Guyana, health workers stated that being able to contact each other by radio telephone greatly improved their morale (Goldschmidt, Forsythe, & Hudson, 1980). Also, in Lesotho, rural field nurses were adamant that if the government expected health staff to accept assignments in more isolated regions, it should provide a means for them to communicate with each other.[12]

TARGETING PRIORITY PROBLEMS

It is important that telemedicine projects be designed to respond to priority medical problems in developing regions. For example, in subsaharan Africa, infant mortality, AIDS, and malaria are among the most prevalent health problems. A focus on prevention, with emphasis on campaigns targeted at high-risk groups, may be the most effective strategy. ICTs could be used to train community health workers, disseminate information to target populations, and coordinate outreach activities.

As noted earlier, telemedicine can help to address the problem of lack of physicians by providing links between rural paraprofessionals and physicians at the district or regional level, thus making scarce medical expertise more widely available throughout the country. Although telemedicine projects focusing on interesting but secondary medical problems may be appealing to researchers, they are not likely to be perceived as a high priority for support by the Ministry of Health or other funders following the pilot period. Yet many telemedicine projects target exotic or rare medical problems, or use equipment that is too complex or expensive to be sustainable.

THE ALLURE OF BANDWIDTH

To some practitioners, telemedicine is synonymous with high resolution graphics and motion video for consultation with distant specialists. For example, the Japanese government initiated "Telemedicine Demonstration Tests" linking Kyushu University and the University of Occupational and Environmental Health, Japan utilizing fiber optic communications. It highlighted the technology-driven rationale for such projects: "This experiment hopes to help Japan begin a full-scale broadband telemedicine test using fiber optics and showcase concrete examples of the advanced information society by developing the usage of the system."[13] The trial was designed to diagnose intractable epileptics by exchanging, via a high speed digital circuit, multichannel digital brain wave signals and synchronized video of patients experiencing an epileptic seizure. To assess the viability of such a telemedicine experiment, evaluation should have included analysis of whether this problem was a high medical priority, how much bandwidth was actually required, and in this case, whether the researchers could actually transmit data while the patient was having a seizure, and whether real-time analysis was critical.

The Japanese universities planned a connection with the Cleveland Clinic Foundation in Ohio, described as "the world's first full-scale international broadband application test of its kind, which will contribute to the implementation of the Global Information Infrastructure (GII) and the information-networked society. . . . Furthermore, both universities hope to be active players in providing telemedicine applications throughout Asia."[14] Yet, it was not stated whether these institutions would be able to pay for the bandwidth after the trial period, whether they had considered licensing and legal issues that would allow them to diagnose and treat patients from other countries, rather than to simply exchange medical information and research, and whether the proposed expansion to other parts of Asia addressed priority medical problems in these regions.

A European Union summary on telemedicine projects stated: "The quality of images limits the full development of telemedicine." This mentality is also seen in some developing country telemedicine projects and proposals. However, rural areas of the developing world typically do not have infrastructure with sufficient bandwidth for such services. Where such capacity may be available (such as fiber optic links between regional centers), the cost of transmission and of onsite facilities may be prohibitive. A simple low-cost add-on to individual telephones is the speaker phone; both the patient and the health worker, or groups such as expectant mothers, can listen in and participate in a dialogue. Audio consultations and audio conferencing can be highly effective, and useful patient data such as chest sounds, EKG tracings, and medical records may also be sent over a telephone line. Digital photos can also be sent as e-mail attachments, giving doctors a visual image of the problem (such as a wound, fracture, or rash).

Some use has been made of full motion video for consultation, but this has not been cost effective, although lower cost digital "compressed video" systems are now available that use a fraction of the bandwidth of conventional analog video systems. Planners should determine how much bandwidth is actually required, and whether real-time analysis is critical (otherwise slower "store-and-forward" data transmission or even mailing of tapes or CD-ROMs may suffice). If video is required, it is important to determine how often it may be used and what quality of image is required. For example, true color reproduction may be critical for some dermatological diagnoses, while high resolution may be necessary for reliable x-ray imaging.

THE INSTITUTIONAL CONTEXT

Investment in technology alone will not likely result in major health benefits. Numerous telemedicine projects were carried out in North America in the 1970s and 1980s, yet despite evaluations that found that in general the technology used achieved the identified goals for diagnosis, consultation, or education, most projects did not survive past the pilot phase (Higgins, Dunn, & Conrath, 1984; Rockoff, 1975). Several factors are cited by various authors:

- Resistance from doctors who feel threatened by new technologies or procedures;
- Failure to integrate the telemedicine system with the organizational structure and culture of the health care system;

- High initial expense of setting up some telemedicine systems, and difficulty in justifying high costs when funds are scarce for higher priorities;
- Lack of funds for ongoing operating costs of telemedical equipment, communication services and technical support;
- Workload issues: regional physicians and urban specialists may be overloaded with local demands on their time and expertise, and therefore unwilling or unable to provide teleconsultations;
- Reimbursement issues: insurers may not reimburse providers who provide teleconsultations;
- Legal liability issues: consultants may be unwilling to provide advice about patients they cannot examine because of potential legal liability.

These factors still limit the utilization of ICTs in health care, such as for electronic patient records and distant consultation.

The strategy of tracking who pays and who benefits can be particularly illuminating for telemedicine projects. In many instances, the health care system pays, but the patient alone benefits. Although this may appear to be sound social policy, if cost savings cited by telemedicine advocates accrue to patients only rather than to the health care system (e.g., avoiding personal transportation to the city for medical tests, accommodation expenses, lost days of work, etc.), there is little incentive for the health care provider to keep paying for the technology.

Escalating medical costs and the shortage of rural physicians are creating incentives for the health care sector to resolve some of the issues that have hindered telemedicine implementation. For example, in the United States, the federal government and several states now reimburse for teleconsultation.[15] Issues such as whether physicians will be paid for teleconsultations are important in a fee-for-service system, but in a government operated system they may not be relevant. However, there may be another problem: Savings in personnel time for consultations and administration may not be an important issue. If the health care system is highly bureaucratic and lacks incentives for cost controls, it has little incentive to introduce cost-effective telemedicine applications.

Telemedicine in Alaska

Alaskan villages are served by community health aides, local residents (primarily women) who receive basic medical training and provide first line care for the villagers. Health aides are supervised by medical staff in regional hospitals that are now operated by native health corporations with funding from the U.S. Public Health Service, which is responsible for providing health care for native Americans.

The health aide system was established in 1954, when a U.S. government report stated that "the indigenous peoples of Native Alaska are the victims of sickness, crippling conditions, and premature death in a degree exceeded in few parts of the world" (Nice & Johnson, 1999). The program began with training of sanitation aides who returned to their villages to instruct others in maintaining safe drinking water and proper trash disposal. In 1956, a physician based in Bethel made the case for expanding the program to train community aides as firstline health workers: "It is not a question of whether the villagers shall be treated by completely qualified medical personnel or persons with less than full qualifications, but a question of whether they shall be tested by persons with limited qualifications or go untreated altogether."

Health aide Paula Ayunerak described those early days: "We had no clinic. We went from house to house taking care of the sick. . . . Our tools consisted of a thermometer, a stethoscope, and a blood pressure cuff. . . . We had no phones, no radios, but used the school's radio to report our patients. There was no nonsense about confidentiality" (Nice & Johnson, 1999). Described as "a marriage between necessity and innovation," the Community Health Aide Program (CHAP) continued to expand. In 1972, villages in central Alaska began to communicate with the Tanana regional hospital and the Anchorage Native Medical Center using a single channel on NASA's ATS-1 satellite. The experiment showed that reliable communications could indeed save time and even lives, and that health aides also learned from each other's experiences heard in consultations over the shared audio channel (see Hudson & Parker, 1973). As a result, the State authorized an expenditure of $5 million for the purchase of satellite earth stations for 200 villages, that communicated through RCA's first commercial domestic satellite.

Today, community health aides are still the frontline providers of village health care, but there are initiatives underway to extend their capabilities through telemedicine. A project known as AFHCAN (Alaska Federal Health Care Access Network) is extending the use of the Internet for telemedicine throughout rural Alaska, using the satellite network. The equipment is designed to be simple and cost effective, consisting of a personal computer and peripherals including EKG (for heart monitoring), electronic otoscope for observing otitis media (ear infections common in village babies), and a digital camera to send still images of patient wounds, lesions, etc. Much of this communication can be done in store-and-forward mode; for example, the digital photos can be sent as e-mail attachments:[16]

> Imagine turning on your computer, clicking onto e-mail, and finding a message from a Community Health Aide . . . at the St. George clinic [in the Pribilof Islands, 770 miles or about 1200 km from Anchorage]. The e-mail message has an attachment that contains heart sounds, which you can actually listen to through your computer or it could contain a picture of a damaged finger or an inflamed eardrum. As a physician based in Anchorage, you would be able to

help the [health aide] down at St. George quickly assess the health problems of a patient thousands of miles away.[17]

AFHCAN may now be the world's largest telemedicine project, serving more than 235 sites, including 198 native clinics.[18]

HealthNet: Medical Links in Cyberspace

SatelLife is a Boston-based nonprofit organization operating HealthNet, a computer network providing communications services and medical information to health care workers in the developing world. SatelLife was founded as a "broker of socially responsible connectivity between health care workers and sources of medical information" to provide low cost access to medical information for health workers in countries where telecommunications facilities were either unavailable or unaffordable. Using a single low earth orbiting (LEO) satellite, HealthNet provided e-mail and document exchange services on a store-and-forward basis using inexpensive satellite terminals.

One of the functions of HealthNet is to provide access to medical information that is not available in underfunded developing world medical schools. For example, while a U.S. medical library subscribes to about 5000 journals, the Nairobi University Medical School Library, regarded as a flagship center for medical literature in East Africa, receives only 20 journal titles today compared to 300 titles a decade ago. A large district hospital in Brazzaville, served by 20 doctors, had a library consisting of a single bookshelf, while the University hospital had only 40 outdated books and a dozen medical journals, none more recent than 1992. Using HealthNet, medical staff can access BITNIS (Batch Internet National Library of Medicine Information System) to conduct electronic searches of data bases at the Washington, DC-based National Library of Medicine (NLM). Selected articles can then be transmitted and downloaded via satellite.

HealthNet also enables medical professionals in developing regions to share information and seek assistance. In early 1995, physicians in central Africa shared vital, up-to-the-minute information over HealthNet during the outbreak of the deadly Ebola virus. In *The Coming Plague: Newly Emerging Diseases in a World Out of Balance*, Laurie Garrett notes: "For the first time, physicians in the developing countries could consult colleagues in neighboring nations or medical libraries and data banks to help solve puzzling cases and alert one another to disease outbreaks." The Muhimbili Medical Center in Dar es Salaam, Tanzania, has turned to cyberspace to help reduce the high mortality rates among its pediatric burn patients. Through HealthNet, the Health Foundation in New York learned of the center's needs and responded by sending a free shipment of phenytoin, a drug that reduces pain and promotes healing of burn wounds.

SatelLife has also evolved to become an Internet gateway, providing public health and environmental workers with an inexpensive "on-ramp" to the global information superhighway. The HealthNet user composes the message or research request on a computer which sends it to a national node that collects, forwards, and receives electronic messages and transmits them to the Internet four times daily. Where Internet access is now available, HealthNet's "store and forward" technology allows messages to be composed off line and stored for later delivery, avoiding the high charges conducting research or composing messages on line. For physicians and researchers working in rural and isolated areas, SatelLife has built a portable ground station; it has also designed a two-way radio unit that will provide users with a seamless interface to the Internet.

NOTES

1. See www.healthnet.org
2. Rosser, J. C. Jr., Gabriel, N., Herman, B., Murayama, M. "Telementoring and Teleproctoring." World Journal of Surgery, 2001, Nov. 25(11):1438–48.
3. See www.orbis.org
4. See www.amref.org
5. See http://www.med.mun.ca/telemed/telemedi.htm
6. See www.healthnet.org
7. See http://www.med.mun.ca/telemed/telemedi.htm
8. See, for example, www.psi.org and www.aed.org
9. See www.healthnet.org
10. See www.amref.org
11. See www.healthnet.org
12. See www.amref.org
13. See Japanese Ministry of Internal Affairs and Communications, www.soumu.go.jp/joho_tsusin/eng/index.html
14. See note 13.
15. See Office for the Advancement of Telehealth http://telehealth.hrsa.gov
16. See www.afhcan.org
17. Carlsson, Sylvia. "Telemedicine in the Bush: A Glimpse into the New Century." www.telemedicine.alaska.edu
18. See www.afhcan.org

Chapter 5

Digital Divides:
Gaps in Connectivity

The Information Highway . . . is [rather] a personalized village square where people eliminate the barriers of time and distance and interact in a kaleidoscope of different ways.
—Information Highway Advisory Council (1994)

The following four chapters examine the problem of increasing access to telecommunications services including the Internet in rural and developing regions. This chapter examines the concept of the "digital divide" and various initiatives, typically in the form of pilot projects and training activities, that have been implemented to address this problem. Chapter 6 examines the broader concepts of universal service and access, and policy initiatives that could provide long-term solutions to gaps in access to connectivity. Chapter 7 reviews technological options and innovations that could help to bridge digital divides. Chapter 8 addresses restructuring strategies that could create incentives for increased investment in ICTs in developing regions.

THE CURRENT STATE OF CONNECTIVITY

During the past 15 years, there has been dramatic growth in global availability of ICTs, from basic telephony to the Internet (see Table 5.1). Growth rates were highest in newly introduced technologies and services, such as cellular telephony and the Internet. By 2003, there were more mobile telephone subscribers globally than fixed lines.

TABLE 5.1
Global Access to ICTs: 1991–2003

	1991	2003	Percentage Increase
Main telephone lines	546 million	1.21 billion	222
Mobile cellular subscribers	16 million	1.33 billion	8,306
Personal computers	130 million	650 million	500
Internet users	4.4 million	665 million	15,045

Source. Derived from International Telecommunication Union (2003).

In industrialized countries and other high-income countries, telephone service is almost universally available through fixed lines and increasingly through wireless networks. Access to voice services has also improved dramatically in developing countries, thanks largely to newly available and more affordable wireless (cellular) services. However, in developing countries both income and location still divide telecommunications haves and have-nots. Poorer countries have fewer telephone subscribers per 100, and within poorer countries, there are greater disparities between urban and rural access than in wealthier countries.

Also, when it comes to Internet access, the news is not good. The term *digital divide* was coined in the 1990s to describe the gap between ICT "haves" and "have-nots" (i.e., those with and without access to telecommunications and Internet services). Digital divides exist within industrialized countries, with the Internet generally less accessible to low-income and disabled populations and to ethnic minorities. In rural areas, those who are connected typically pay more than their urban counterparts for Internet access. Disparities in Internet access are found in the Canadian North and the Australian Outback, and in rural and disadvantaged parts of Europe. But the divide is a chasm in the developing world—between industrialized and developing countries, and within developing countries between the privileged, predominantly in urban areas, and the disadvantaged, who are often rural as well as poor. Africa accounts for about 13.5 percent of the world's population, but only 1.8 percent of the world's Internet users (see Table 5.2).

FROM NII TO GII

The phenomenal growth and visibility of the Internet as an information resource, communications tool, and electronic marketplace has focused attention on the need for investment in telecommunications facilities to bring the Internet and other forms of electronic communications within reach of people around the world. In 1993, the newly elected Clinton administration called for investment in national information infrastructure

TABLE 5.2
Internet Access by Region

	Internet Users (in millions)	Global Percentage of Internet Users	Region's Percentage of Global Population
Africa	12.8	1.8	13.5
Americas	224.5	32.4	13.9
Asia	249.9	36.0	59.1
Europe	192.6	27.8	13.0
Oceania	13.7	2.0	0.5

Derived from ITU ICT statistics, 2005.

(NII) to bring the benefits of "advanced services," and particularly the Internet, to Americans not only to the workplace, but to schools, libraries, health care centers, and to individual households. Many other industrialized countries were soon spurred to action by the astonishing growth of the Internet and concern about the economic and technological lead the United States might gain through its commitment to NII.[1] Notably, the Bangemann Report to the European Union stated: "The first countries to enter the information era will be in a position to dictate the course of future developments to the late-comers."[2] (In fact, Singapore may have been the first to implement an information age strategy much earlier through its commitment to becoming a regional information hub and "intelligent island.") Many countries took up the task of developing NII plans, most of which were remarkably similar. Most recommended government support for II trials and projects. Some set specific targets such as Japan's fiber to all homes by 2010, Singapore's *IT 2000* and Malaysia's *Vision 2020*.[3]

U.S. Vice President Al Gore elaborated on this theme at the ITU's Development Conference in Buenos Aires in 1994, calling for a Global Information Infrastructure (GII) to extend access to these new technologies and services to people throughout the developing world. Al Gore caught the imagination of traditional telecommunications administrators through his ITU speeches; building on this introduction, the ITU's Telecom 95 Exhibition included for the first time sessions on the Internet and computer and software vendors as well as the traditional telecommunications carriers and equipment suppliers.

However, it was also important to reach national decision makers and not just the technocrats. A major event aimed at industrialized country leaders was the Information Society Showcase, held in conjunction with the G7[4] Ministerial Meeting in Brussels in February 1995, where G7 ministers and other senior officials got hands-on opportunities to share lessons with students around the globe in a "world classroom"; use a digital map that could be instantly updated over a mobile phone; use a computer that could read electronic mail out loud; work in two offices in different places at the same time; and turn themselves into cartoon characters. Surely something there to capture the attention of any politician!

Thus, in 1995, the G7 nations together embraced the goal of global information infrastructure, and initiated a series of demonstrations and pilot projects using high capacity networks and switching as a means of demonstrating the potential benefits of ICTs. In Japan, the Nippon Telegraph and Telephone Corporation (NTT) announced its intent to wire every school, home, and office with fiber optic cable by the year 2010. Japan's Ministry of Posts and Telecommunications (MPT) estimated the cost of building this network to be between $150 billion and $230 billion. The Japanese also funded projects and trials such as a $50 million 3-year pilot project to assess the feasibility of integrating telecommunications and broadcast services such as video on demand, high-definition television, videoconferencing, teleshopping, and telemedicine through fiber-to-the-home networks.[5]

The APEC (Asia Pacific Economic Cooperation[6]) nations also sponsored projects using telecommunications networks for distance education and training, health care delivery, and economic development. APEC members include not only the industrialized economies of Japan, Singapore, the United States, and Canada, but also countries with much greater gaps in their infrastructure such as China, Thailand, Indonesia, and the Philippines. Developing country leaders around the world were beginning to recognize the importance of ICTs and the danger of falling further behind in gaining access to the Internet. The 1996 Information Society and Development (ISAD) Conference held in South Africa was the first international conference convened in a developing country to consider the potential of the new information and communication technologies (ICTs) for development. ISAD noted a link between the information society and community development in Africa and suggested that, "through links at the community level, a Global Information Community governed by people-centred development values could evolve within the more commercially oriented Global Information Society." Meanwhile, African governments had begun to recognize the potential contribution of ICTs to African development, and had established an African Information Society Initiative (AISI).

The World Bank added its imprimatur by cosponsoring the first Global Knowledge Conference in 1997 in Toronto. The initiative for the conference was spearheaded by World Bank President James Wolfensohn, an almost evangelical convert to ICTs. More than 2,000 attendees of GK97 could attend sessions on distance education, telemedicine, and other developmental applications of ICTs, as well as demonstrations and electronic testimonials. An Amazon tribal chief explained via teleconference how his village had struck a deal with British retailer Body Shop, to sell essential oils from the rain forest.[7] Inuit children from the Canadian Arctic demon-

strated their website, where they published videos of elders teaching traditional crafts and exchanged letters with children in other countries (Wysocki, 1997).

At its summit in Okinawa, Japan, in July 2000, the members of the G8
(which now included Russia) endorsed a declaration known as the Okinawa Charter on the Global Information Society, which stated:

> The essence of the IT-driven economic and social transformation is its power to
> help individuals and societies to use knowledge and ideas. Our vision of an in-
> formation society is one that better enables people to fulfil their potential and
> realise their aspirations. To this end we must ensure that IT serves the mutually
> supportive goals of creating sustainable economic growth, enhancing the pub-
> lic welfare, and fostering social cohesion, and work to fully realise its potential
> to strengthen democracy, increase transparency and accountability in gover-
> nance, promote human rights, enhance cultural diversity, and to foster interna-
> tional peace and stability. . . . In pursuing these objectives, we renew our com-
> mitment to the principle of inclusion: everyone, everywhere should be enabled
> to participate in and no one should be excluded from the benefits of the global
> information society.[8]

The Digital Opportunities Task Force (DOT Force) was established to
recommend strategies for implementation of the Okinawa Charter. It
brought together 43 teams from government, the private sector, nonprofit
organizations, and international organizations, representing both developed and developing countries, in a cooperative effort "to identify ways in
which the digital revolution can benefit all the world's people, especially
the poorest and most marginalized groups." The DOT Force presented its
report to G8 leaders in Italy in 2001, where it put forward a plan of action
(the Genoa Plan of Action) intended to provide a basis for developing
economies to achieve sustainable social and economic development enabled by ICTs. The plan also provided a framework for possible initiatives
by member governments and other stakeholders, including a call for initiatives that would:

- Help establish and support developing country and emerging economy national e-strategies;
- Improve connectivity, increase access and lower costs;
- Enhance human capacity development, knowledge, creation and sharing;
- Foster enterprise and entrepreneurship for sustainable economic development;
- Establish and support universal participation in addressing new international policy and technical issues raised by the Internet and information and communications technologies;

- Establish and support dedicated initiatives for the information and communications technologies inclusion of less-developed countries;
- Promote information and communications technologies to support health care and the fight against HIV/AIDS and other infectious and communicable diseases;
- Encourage national and international efforts to support local content and applications creation;
- Prioritize information and communications technologies in the G8, as well as other development assistance policies and programs; and
- Enhance the coordination of multilateral initiatives.[9]

The DOT Force presented a progress report to the G8 leaders at their next summit in Canada in 2002, after which it ceased to exist. The intent was that other international and multilateral bodies, such as the United Nations, would carry on implementation of the initiatives approved by the G8 leaders. The United Nations established the UNICT Task Force, with members representing governments, civil society (including the private sector, foundations, NGOs, and academia) and UN agencies.[10]

DIGITAL DIVIDE INITIATIVES

Numerous international organizations and development agencies have taken up the cause of the digital divide. International organizations that have specific initiatives concerned with Internet access and applications include the ITU, UNESCO, UNDP and the World Bank. Other sector-specific agencies such as FAO and WHO also have funded projects to apply information technologies including the Internet. Examples include:

- **UNDP's Sustainable Development Networking Program:** The Sustainable Development Networking Programme (SDNP), sponsored by the United Nations Development Program (UNDP), was a national information exchange operation run by independent entrepreneurs who receive equipment and seed funding from UNDP. It operated at the country level, launching and supporting local Internet sites, and building national capacities and knowledge resources.[11] A direct result of the 1992 UN Conference on the Environment and Development (the Rio Earth Summit), the SDNP linked government organizations, the private sector, universities, non-governmental organizations (NGOs) and individuals through electronic and other networking vehicles to exchange critical information on sustainable development. SDNP operated in 24 countries; another 75 countries expressed interest in establishing networks. It was not limited to the Internet, but supported use of a wide variety of communication technologies, including radio and audio cassettes (Lankester & Labelle, 1997).

• **The World Bank** funds telecommunications infrastructure projects, often as a lender of last resort, supplementing funds from traditional financial and industry sources. It also invests as a minor partner in some private sector joint ventures such as cellular networks. Its recent sector strategy changes its mission from expanding and modernizing telecommunications through investment in fixed and mobile voice and data networks to extending access to a wider range of ICTs and related applications, and investing in Internet and broadband networks and sector-based applications.[12] In the 1990s, the World Bank also established new initiatives specifically to fund activities using ICTs for development rather than infrastructure alone, including the following:

InfoDev: The World Bank points out that ICTs "open up extraordinary opportunities to accelerate social and economic development, and they create a pressing reform and investment agenda both to capitalize on the new opportunities and to avoid the deterioration of international competitiveness."[13] Its Information for Development Program (InfoDev) aimed to address this agenda by funding activities to assist developing countries and emerging economies to harness these technologies. Its strategies included leveraging funds and brokering partnerships to create a network for improved communication and information sharing.
InfoDev's main objectives were to:

- Create market-friendly environments;
- Reduce poverty and exclusion of low-income countries and social groups;
- Improve education and health;
- Promote protection of the environment and natural resources;
- Increase the efficiency, accountability and transparency of governments.[14]

It funded numerous pilot projects and short activities such as training courses and workshops. In 2004, InfoDev changed its strategy to focus on the linkages between pilot projects, evidence, analysis, and action in harnessing ICTs for development. Accordingly, InfoDev planned to launch "an intensive program of support for research, analysis, and evaluation, impact monitoring, and toolkit development focused on distilling the lessons of experience from the past ten years on the impact of ICT on poverty, with a particular focus on mainstreaming and scaling up successful ICT approaches and applications."[15]

World Links: The World Bank also established World Links for Development in 1997 as a pilot initiative in response to widespread requests from developing countries to assist them in preparing their youth to enter an information age. World Links focused on providing Internet access for schools in developing countries. Reaching more than 200,000 teachers and students in 22 developing countries, World Links supported mobilization of "the equipment, training, educational resources and school-to-school, NGO and public–private sector partnerships required to bring students in developing countries online and into the global community."[16] World Links was spun off as an independent NGO after its five-year World Bank pilot.[17]

Development Gateway: The World Bank established the Development Gateway as a web portal on development issues, from which users can access information, resources, and tools, and to which they can contribute their own knowledge and experience. The Gateway is intended to "enable those in the development field to share information, easily communicate, and build communities of practice around significant challenges from the grassroots up." It is now operated by the nonprofit Development Gateway Foundation.[18] While intended as a resource on development issues in general, and not specifically on connectivity, it demonstrates the power of the Internet to be a tool to access and share information for development. Other organizations are building their own portals to serve regional needs, such as South African SANGONeT's Africa Pulse portal.[19]

• **The International Telecommunication Union (ITU) and UNESCO** jointly sponsored multipurpose community telecentres (MCTs) in developing regions. In Africa, MCTs were installed in Benin, Mali, Mozambique, Tanzania, and Uganda. The ITU also supported telemedicine projects in Ukraine, Mozambique, Malta, Myanmar, Georgia, Senegal, and Uganda, as well as other countries.[20] The ITU and UNESCO have also collaborated with other specialized UN agencies such as WHO (World Health Organization), FAO (Food and Agriculture Organization), and UNITAR (United Nations Institute for Training and Research).

BILATERAL INITIATIVES

Most bilateral development agencies have prepared a strategy on ICTs for development. Initiatives sponsored by Canada and the United States are highlighted below. Other examples include the United Kingdom's Building Digital Opportunities Programme through its Department for International Development (DFID) and the Japanese Overseas Development

Agency's (ODA) Comprehensive Cooperation Package to Address the International Digital Divide.

• **Canada's International Development Research Centre (IDRC):** IDRC operates a global networking program consisting primarily of the Pan Asian Networking (PAN) initiative which was expanded to cover other developing regions including Latin America, the Caribbean, North Africa, and the Middle East, and the Acacia initiative which addressed the information needs of sub-Saharan Africa. IDRC assists developing country NGOs and entrepreneurs in establishing Internet service providers (ISPs), provides training, and emphasizes sustainability. It has provided startup funding for ISPs in Mongolia, Vietnam, and Bangladesh, and has projects underway in several other developing countries in Asia, Africa, and Latin America.[21]

• **USAID:** The U.S. Agency for International Development (USAID) has sponsored several initiatives designed to extend access to the Internet and to foster development-related applications of ICTs.

Leland Initiative: The Leland Initiative (named for deceased Congressman Mickey Leland), sponsored by USAID was a five-year $15 million U.S. Government effort to extend full Internet connectivity to approximately 20 African countries in order to promote sustainable development. Leland had three major objectives:

- Policies: to create an enabling policy environment;
- Pipes: to foster a sustainable supply of Internet services;
- People: to promote Internet user applications for sustainable development.

More than 20 sub-Saharan African countries participated. Leland expanded its activities to assist African governments in telecommunications policy formation, provide training in ICT applications, and introduce wireless technologies for "last mile" access.[22]

LearnLink: Between 1996 and 2003, USAID funded the LearnLink project, which implemented about 20 ICT-based activities in developing countries. These activities in Africa, Latin America, and Eastern Europe involved sectors ranging from basic education to teacher training, professional development, participant training, lifelong learning, economic development, municipal networking, health, and institutional and organizational strengthening.[23] LearnLink was succeeded by the DOT-COM Alliance, a series of initiatives also funded by USAID focusing on extending access to underserved communities (dot-ORG), strengthening education and learning systems through customized ICT interventions and content (dot-EDU), and promoting policy and regulatory reform (dot-GOV).[24]

FOUNDATION AND CORPORATE INITIATIVES

Among major foundations supporting initiatives to improve access to and effective use of the Internet in developing regions is the Soros Foundation, whose Open Society Institute (OSI) is a private foundation that promotes the development of open societies around the world by operating and/or supporting a variety of initiatives and projects in education, independent media, legal reform and human rights. The OSI's Internet Program (IP) supports projects to provide Internet access as part of its strategy for fostering open civil societies. The program focuses on universities and related research institutions, secondary (and sometimes primary) schools, libraries, medical institutions, cultural centers such as museums and galleries, NGOs, independent media, environmental groups, and unaffiliated individuals who might otherwise not have access to connectivity services.[25] The Benton Foundation and the Markle Foundation both expanded their domestic communication focus to include international activities. Markle funded part of the DOT Force activity, but has apparently since lost interest, changing its focus to health and national security. Benton's online Digital Divide Network has added an international channel for developing country issues.[26]

Major high tech companies (such as Apple, Dell, Hewlett Packard, and IBM) that combined philanthropy with enlightened self-interest for many years by donating or substantially discounting equipment for schools and underprivileged areas of industrialized countries, are now extending their activities to the developing world. Some, such as Cisco Systems and Nortel Networks, emphasize training, either through their own corporate programs or with public sector partners. Cisco's Networking Academy (see chap. 3) launched in 1997 with 64 educational institutions in seven states has expanded to more than 150 countries and all 50 U.S. states. Some 1.6 million students have enrolled at more than 10,000 academies located in high schools, technical schools, colleges, universities, and community organizations.[27] Intel's Teach to the Future is designed to help teachers integrate technology into instruction to develop students' higher level thinking skills and enhance learning. Intel states that the initiative has trained more than two million teachers in 30 countries, with support from Microsoft.[28] HP's E-Inclusion initiative has supported community ICT projects in minority communities in the United States, South Africa, Brazil, and India. HP sees E-Inclusion as means to create new market opportunities for the company and for the communities. For example, in South Africa it has developed a means of enabling four students to work independently sharing a single computer; this product may be introduced in other emerging markets.[29]

The private sector also helps to support several training programs. The United States Telecommunications Training Institute (USTTI) provides short courses in the United States for telecommunications officials from developing countries. Funded by USAID and U.S. telecommunications and high tech companies, USTTI has added training courses on universal service, the Internet, and e-commerce policy issues.[30] Management training for telecommunications professionals is also provided by a Canadian public/private sector initiative called TEMIC[31] and by the Commonwealth Telecommunications Organization (CTO).[32] Telecommunications companies such as Cable and Wireless and NTT also provide training for developing country professionals.

Most of the above donors, except for agencies that fund infrastructure projects such as the World Bank and some bilateral donors, have not invested in connectivity. Instead, they have focused on utilization of the Internet for development goals through training and production of development-related content for schools, nonformal education, environmental monitoring, small business skills and markets, etc. However, some now include support for policy initiatives, having realized that affordable access to computers and the Internet will not be achieved in many developing countries without changes in their governments' policies.

RURAL PROJECTS IN INDUSTRIALIZED COUNTRIES

In industrialized countries, the rhetoric about information infrastructure and information superhighways was accompanied by primarily government-sponsored projects designed to extend services to rural or remote areas, or to demonstrate applications of telecommunications for development-related activities. However, reviews of these projects were generally descriptive rather than analytical, and lacked data that could contribute to a better understanding of costs and benefits. A few are included here to illustrate types of rural telecommunications projects that have been supported in industrialized countries.

Canada

In the 1970s, the Canadian government recognized the potential of satellites to reach the remote North. The purpose of government-sponsored experiments and pilot projects was to contribute to regional development in health and education using satellite technology, which itself was funded by the Canadian government. Little evaluative research was published. However, one pragmatic means of assessing the value of such projects is to determine which were continued. In fact, several projects that originated during the experimental satellite era are now operational programs using commercially available satellites and terrestrial networks. They include Memorial University's telemedicine service in Newfoundland, the Knowledge Network (which became the BC Open University) in western Canada, a radio network run by Cree and Ojibway people in northern Ontario, and a television network run by Inuit Broadcasting Corporation in the Arctic.

Canada then planned to build a network of networks linking Canadian communities, businesses, government agencies, and institutions. The Canadian government saw the information highway, as it was known in the 1990s, as a catalyst to help Canadians share information, and to gain an edge in productivity and information industries in global markets. It identified three key objectives for the Canadian information highway:

- To create jobs through innovation and investment;
- To reinforce Canadian sovereignty and identity;
- To ensure universal access at reasonable cost.[33]

These objectives emphasize cultural as well as economic priorities. Canada's Information Highway Advisory Council, composed of representatives of communications industries, business users, academics, performers, and consumers, took a distinctly cultural perspective:

> The Information Highway, in our view, is not so much about information as it is about communication in both its narrowest and broadest senses. It is not a cold and barren highway with exits and entrances that carry traffic, but a series of culturally rich and dynamic intersecting communities . . . . Rather than a highway, it is a personalized village square where people eliminate the barriers of time and distance and interact in a kaleidoscope of different ways.[34]

In 1998, Canada launched a "Connecting Canadians Initiative" that included support for community access to the Internet, connecting schools and libraries, and increasing access and skills among target groups such as native people, ethnic and linguistic minorities, and remote communities.[35]

European Union

In western Europe, the European Community supported several rural telecommunications initiatives. STAR (Special Telecommunications Action for the Regions) was aimed at peripheral regions, to create conditions and provide incentives for extension of telecommunication services to smaller regions. A major criticism of the STAR initiative was its bias toward infrastructure, which accounted for about 80 percent of total expenditure of the program. The Télématique program was designed to promote use of advanced telecommunication services in regions where development is lagging behind the rest of the Community. The ORA initiative (Opportunities for Rural Areas) was a research and development initiative aimed to identify the potential opportunities offered by the application of "telematic" systems in rural areas, and to identify the technologies and services required throughout the rural areas of the EC to realize these opportunities. In northern Scotland, the Highlands and Islands Initiative was designed to integrate telecommunication policy with regional development policy through upgrading British Telecom's network with digital switching and ISDN capability. It was intended to act as one of a range of incentives encouraging relocations into the region. However, although there were anecdotes about telecommuters including consultants, architects, and telephone operators working from home, no evaluative studies were released (Taylor & Williams, 1990).

Scandinavia

Scandinavia (primarily Denmark and Sweden) originated telecottages, typically small buildings or rooms in rural communities equipped with a few personal computers, printers, modems, and a fax machine. There were two main variants. Community telecottages provided a range of services on a nonprofit basis, and were intended to benefit the local community by providing help to local businesses and farmers. Business telecottages resembled city-based business bureaus and provided a range of services on a profit-making basis. Services commonly available at Nordic telecottages were information retrieval from libraries and databases; consultancy and ITT training; and access to teleworking facilities for local people (Qvortrup, 1989). Again, there was little systematic evaluation of costs and benefits. However, one consistent finding was that an important element for initial usage was the resource person who provided training and other guidance. It appeared that other factors were needed for the project to go beyond the awareness and training phase, such as local entrepreneurs, business experience, and contacts with potential contractors for rural information services such as data entry and telemarketing. These conclusions are relevant for telecenters in developing countries, which were based on the telecottage model.

United States

With the call for Information Superhighways in the 1990s, there was a resurgence of funding of pilot projects in the United States that had not been seen since the federal support of experimental satellite projects in the 1970s. The Rural Electrification Administration (REA), now renamed the Rural Utilities Service (RUS), established a Distance Learning and Medical Grant Program.[36] The U.S. Department of Education funded distance education through its Star Schools Program and other initiatives. The National Telecommunications and Information Administration (NTIA) funded pilot projects and infrastructure grants for the planning and construction of telecommunications networks for educational, health, library, and other social services.[37] Some states sought to combine high-quality infrastructure with attractive rural settings to attract "footloose" entrepreneurs such as consultants, architects, and software developers who appreciate the quality of rural life. (In Colorado, these people are called "lone eagles"; in Montana, they are known as "modem cowboys.") For example, the Colorado Advanced Technology Institute (CATI) provided seed money to selected rural communities that proposed using telecommunications as part of their economic development strategies, and funded rural gateways to its statewide computer network (Richardson, 1993).

CRITICAL ISSUES

Although the phenomenal growth of the Internet and development of Internet-based businesses in the 1990s focused a great deal of attention on the digital divide and resulted in support for numerous projects and other initiatives, such activities may have numerous problems and pitfalls. The following section examines lessons learned and critical issues that need to be addressed in implementing projects and trials of new communications technologies.

Fads and Funding

The initiatives described above and the boom in the dot-com industries in the late 1990s resulted in dramatic increases in attention focused on ICTs in development and in funding for ICT field projects, training and policy studies. The result was too often technology in search of a problem and/or duplication and fragmentation of projects, as development agencies, foundations, and corporate donors jumped on the ICT bandwagon. But with the so-called dot-com bust of the early part of the new decade, corporate funding decreased, and funding agency agendas shifted to more traditional development sector priorities. The ICT fad was passé. ICT projects were particularly vulnerable to trends in the high tech industries. This vulnerability mirrors earlier phases of communication project funding spurred by technological innovations such as the North American experimental satellites and newly introduced cable TV systems of the 1970s.

Technology and Free Tickets

The early phases of the NII/GII movement were primarily driven by the introduction of optical fiber; indeed, in the early 1990s, many pundits predicted the demise of satellites and other wireless systems. In this respect, the Information Superhighway era movement resembled the experimental satellite era of the early 1970s, when satellites were too often a solution in search of a problem, or, in some cases, a solution to a real problem that ignored the many other issues that had to be addressed to solve it. This approach goes beyond "Build it and they will come." It could be called "Build it and give away free tickets." Although this phenomenon is more common in industrialized countries, the rationale of building facilities such as telecenters without understanding local demand is also found in many developing country projects.

Many proponents of NII/GII equated information infrastructure with big pipes and fast switches, such as optical fiber with ATM switching, and later IP networking. Following a technology-driven rationale, they then advocated support for trials and pilot projects that could take advantage of this capacity. Many of the projects in industrialized countries involved research institutions with super computers.[38] Health care trials typically focused on imaging such as CAT scans and MRI images sent between hospitals and specialists at major medical centers. Yet, there was often little effort to identify high priority sector needs, or to determine where ICT interventions were likely to be most useful.

Where Are the Users?

One reason that many ICT initiatives seem overwhelmingly technology driven is that they are implemented primarily by technocrats. For example, in Asia, the Asia-Pacific Telecommunity (APT) identified information infrastructure (APII) projects as a priority in the late 1990s. However, at its High Level Development Meeting on Asia-Pacific Information Infrastructure, not only were there no representatives of users present, but strategies for involving users or identifying priorities for applications were not discussed. Instead, each country, represented by its telecommunications administration, reported on its progress in developing APII initiatives. The regional projects were dominated by Japan and Australia, the major funding partners and sources of expertise. No effort was made to facilitate exchange of experiences among countries with similar problems or shared development agendas, such as the Pacific Island nations or members of ASEAN. Instead, countries simply reported over two days in alphabetical order, with China followed by the tiny Cook Islands, and Kiribati (with a population of 101,000) by South Korea, with more than 48 million people and the world's highest penetration of broadband.[39]

A much more user-inclusive approach was adopted in Hong Kong, where the Office of the Telecommunications Authority (OFTA) established an Information Infrastructure Advisory Committee, with a Taskforce on Applications and working groups on education, business, government, the community, the environment, and personal services.[40] Workshops enabled the public to gain hands-on familiarity with ICTs. The African Information Society Initiative (AISI) also adopted a more inclusive approach, involving stakeholders from civil society, the private sector, academia and African professionals at home and abroad.[41]

Funding Drives Applications

Even when users are involved, the availability of funds for NII/GII and digital divide projects can skew priorities. As was true with the experimental satellite projects of the 1970s and 1980s, projects may be proposed because funds are available, rather than because they may solve a high priority problem such as reducing air pollution in Asian cities or upgrading the skills of rural teachers and health care workers in Africa. An offer of free hammers tends to generate proposals for projects that use a lot of nails, even if nails are not really needed. As with earlier technology initiatives, funding for ICTs by zealous donors ignorant of the local context once again became a solution in search of a problem.

SUSTAINABILITY

Without incentives and links to high priority problems, ICT projects and trials tend to terminate when funding ends or the free ride on the network is over, even if they have demonstrated successful applications of the technology. This problem is not limited to developing countries; there were numerous examples from many of the early North American satellite experiments. For example, in 1977, Dr. Maxine Rockoff, who had funded several U.S. satellite telemedicine experiments through the National Center for Health Services Research, pointed out that successful telemedicine trials were not being continued, even if they proved beneficial in saving time and money, apparently because the U.S. health care system at that time had no incentive to reduce the number of referrals from general practitioners to specialists or to fill fewer hospital beds (Rockoff, cited in Hudson, 1990). Adoption of ICTs by health care systems in industrialized countries is still hindered by institutional barriers such as decentralized records, billing practices, and geographical boundaries on certification. Rather than starting with technology, a more developmentally sound approach would be to examine a nation's or region's social and economic priorities (typically prioritized in government policy documents) and then determine where applications of information technology could significantly contribute to achieving the goal or addressing the problem.

Another flaw in many ICT pilot projects has been that there was no planning or transition strategy to continue the application after the experimental period. Again, this is not a new problem. Most of the early North American satellite projects died when the experimental period was over. Two exceptions, where commitment was coupled with successful entrepreneurial and policy strategies, were the ATS-1 health network in Alaska which led to satellite-based telephone service for Alaskan villages, and the Appalachian ATS-6 project which eventually evolved into the Learning Channel (Hudson, 1990). The more recent information infrastructure and digital divide experiments and trials use operational networks, typically built by telecommunications carriers, although some may use dedicated government networks. Thus, participants do not have to look around for other capacity, but they do have to find the funding to pay for use of the network and continued operation and maintenance of the equipment they need for the application. ICT use can grind to a halt if a school has no funds to repair computers or a telecenter has no money for a new printer cartridge.

USAID's Leland Initiative

The Leland Initiative (LI), sponsored by the U.S. Agency for International Development (USAID), was a 5-year, 15-million-dollar U.S. Government effort to extend full Internet connectivity to approximately 20 African countries in order to promote sustainable development. LI (named for deceased Congressman Mickey Leland) had three major objectives:

- Policies: to create an enabling policy environment;
- Pipes: to foster a sustainable supply of Internet services;

- People: to promote Internet user applications for sustainable development.

Mali was the first country to open a Leland Gateway in December 1996. Private ISPs began to operate with Leland support in Madagascar and Mozambique in 1997. Other countries participating in Leland include: Benin, Botswana, Côte d'Ivoire, Eritrea, Ethiopia, Ghana, Guinea, Guinea Bissau, Kenya, Malawi, Namibia, Rwanda, Senegal, South Africa, Tanzania, Uganda, Zambia, and Zimbabwe.

In signing accords with Leland, the countries agreed to adopt the following "Internet-friendly" policies:

- Abandon traditional international telephone pricing for cost-based affordable tariffs;
- Allow free and open access to the information on the Internet; and
- Set aside long-standing monopoly practices in favor of private sector Internet Service Providers (ISPs).

In return, USAID agreed to provide the policy analyses, equipment, connectivity, and training necessary to deliver robust, well-managed and responsive Internet access. In addition, the Leland Initiative prepared several pilot projects to demonstrate the application of Internet connectivity for ag-business trade information centers, environmental trade links, environmental education and communication, improved local governance, and basic education.[42]

Canada's Smart Communities

The Canadian government initiated a program "to help establish world-class Smart Communities across the country so that Canadians can fully realize the benefits that information and communication technologies have to offer." It defined a Smart Community as "a community with a vision of the future that involves the use of information and communication technologies in new and innovative ways to empower its residents, institutions and regions as a whole."[43] They were to become centres of expertise in the integration of information and communication technologies into communities, organizations and families. The Program's objectives were:

- to assist communities in developing and implementing sustainable Smart Communities strategies;
- to create opportunities for learning through the sharing among communities of Smart activities, experiences and lessons learned;

- to provide new business opportunities, domestically and internationally, for Canadian companies developing and delivering information and communication technology applications and services.

K-Net, The Kuh-ke-nah Network Of Smart First Nations, is a Smart Community project serving remote Ojibway and Cree native communities in Northwestern Ontario, Canada. K-Net is a regional broadband network linking First Nations and their service organizations using a variety of ICTs including video conferencing, IP telephony, on-line forums, e-mail, and other web-based communication tools. Other revenue-generating services that K-Net provides are:

- Computer maintenance and support for the Chiefs Council and First Nations;
- Operating and managing a small computer business;
- Regional hardware and software helpdesk service for Industry Canada First Nations Schoolnet;
- Developing and facilitating computer training programs.[44]

As a result of its experience in establishing telecommunications networks in northern native communities, K-Net offered several observations and suggestions for management and outreach in isolated communities:

- Community involvement: "Include and involve members of isolated communities as early as possible when designing and implementing programs or plans to provide telecommunications services in those communities. . . ." "Getting communities involved in the planning and implementation of broadband connectivity solutions and applications supports local innovation and capacity building. These communities will be better prepared to be the providers of online services and resources 'making them owners of their local networks and producers of local socio-economic opportunities.' "
- Documentation: "Highlight and recognize the value of unique technology applications and solutions in isolated communities. These represent potential opportunities for other communities in Canada and internationally that will otherwise be missed."
- Adaptation: "Processes that are designed in an urban context often include 'macro' entities in a corporate, institutional, or government perspective. In small, isolated communities these same kinds of entities exist at the individual level. Adapt the processes accordingly."
- Visibility: "Leaders of isolated communities should invite government and corporate leaders involved in telecommunications to visit their communities." A visit by senior government and telecommunications officials . . . "was very helpful for them to gain a better appreciation of the physical realities we face here, which in turn benefits us."[45,46]

NOTES

1. As presented by the Clinton Administration in the 1994 NTIA report. *The National Information Infrastructure: Agenda for Action.* Washington, DC: U.S. Department of Commerce.

2. High-Level Group on the Information Society. *Europe and the Global Information Society: Recommendations to the European Council* (The Bangemann Report). Brussels: European Commission, 1994.

3. Their strategies for extending or upgrading infrastructure, however, vary widely from Singapore's state investment in Singapore ONE, to government supported high tech zones such as Malaysia's Multimedia Supercorridor (MSC) and the Philippines' Subic Cybercity, to market-driven competition policies in Hong Kong, New Zealand, and the United States.

4. G7: An association of the world's major industrialized economies: Canada, France, Germany, Italy, Japan, the United Kingdom, the United States. Now includes Russia and is known as the G8.

5. Industry Canada, 1994.

6. Asia Pacific Economic Cooperation, an association of Asian and Pacific enonomies (the use of latter term, rather than nations, allows inclusion of entities such as Hong Kong and Taiwan).

7. See www.idrc.ca/acacia

8. See www.g8.utoronto.ca/summit/2000okinawa/gis.htm

9. Industry Canada. "Canada Chairs Information and Communication Technologies for Developing Meeting." See www.ic.gc.ca

10. United Nations ICT Task Force. www.unicttaskforce.org

11. See www.sdnp.undp.org

12. "Information and Communication Technologies: A World Bank Group Strategy." See http://info.worldbank.org/ict/assets/docs/ExecSum.pdf

13. See www.infodev.org

14. See endnote 13.

15. See endnote 13.

16. See www.worldlinks.org

17. See endnote 16.

18. See www.developmentgateway.org

19. See www.sangonet.org.za

20. See http://www.itu.int/ITU-D

21. See www.idrc.ca

22. Leland Initiative. www.info.usaid.gov/leland

23. See http.learnlink.aed.org

24. See www.dot-com-alliance.org

25. See www.soros.org/initiatives/information

26. See www.digitaldividenetwork.org

27. See http://www.cisco.com/us/learning/netacad/academy/index.html

28. See www97.intel.com/education/teach/

29. See http://www.hp.com/e-inclusion/en/

30. See www.ustti.org

31. See www.temic.ca

32. See www.cto.int

33. Industry Canada (1994). The Canadian Information Highway. Ottawa: Industry Canada.

34. Information Highway Advisory Council (1994). *Canada's Information Highway: Providing New Dimensions for Learning, Creativity and Entrepreneurship.* Ottawa: Industry Canada.

35. See www.ic.gc.ca

36. See www.usda.gov/rus/telecom/index.htm

37. See http://www.ntia.doc.gov

38. For example, Pacific Bell's CalREN (California Research Network) and Canadian University projects sponsored by CANARIE.

39. The author attended the Bangkok APT meeting on AII in June 1997 as an observer.

40. See www.ofta.gov.hk/index_engl.html

41. See www.uneca.org/aisi/

42. USAID Bureau for Africa, Office of Sustainable Development. "Leland Initiative: Africa GII Gateway Project." www.usaid.gov/leland

43. See http://smartcommunities.ic.gc.ca/program_e.asp

44. See http://services.knet.ca

45. See http://smartcommunities.ic.gc.ca/lessons/lessons-Knet2_e.asp

46. For additional background information, see http://smart.knet.ca and http://knet.ca/portal

Toward Universal Access: Strategies for Bridging Digital Divides

[The E-Rate] should be a great equalizer. What we can do is make sure that the inequalities stop at the schoolhouse door and at the front steps of the library. And that is the power of technology.

—Reed Hundt (1997)

The previous chapter examined the concept of the digital divide and various initiatives, typically in the form of pilot projects and trials, that have been implemented as steps to bridge the divide. This chapter examines the broader concepts of universal service and access, and policy initiatives that could provide long-term solutions to gaps in connectivity.

CHANGING CONCEPTS OF UNIVERSAL SERVICE

The term *universal service* originated not as a public policy goal, but as an industrial strategy. Theodore Vail, the early visionary head of AT&T, coined the term universal service to mean the delivery of all telephone services through one network, the Bell system (Mueller, 1997). Vail was later willing to submit to government regulation to achieve his goal, a single, unified carrier providing both local and long distance service throughout the United States. However, by 1920, only 35 percent of American households had a telephone (Crandall & Waverman, 2000). The Communications Act of 1934, which established the Federal Communications Commission, made no specific reference to universal service, although it did invoke "the public interest" and stated that its purpose was "to make available, so far as possible to all people of the United States, a rapid, efficient, nationwide . . . communications service with adequate facilities at reasonable charges." Over the next 40 years, various methods were adopted to apportion the costs of providing telephone service among various elements in the network, which generally resulted in keeping local service prices low through cross-subsidies from long distance and other services.

The term universal service was reintroduced in 1975 in testimony by AT&T consultant (and former advisor to President Lyndon Johnson) Eugene Rostow, who sought to ward off competition from MCI and other new long distance competitors, and to resist the efforts of the Department of Justice to break up AT&T. As Mueller (1997) notes, "In this struggle, the concept of universal service was redefined in a way that linked it to the practices of regulated monopoly." Although AT&T eventually lost the argument that the way to keep local rates low and affordable was to prevent competition, a system of access charges was implemented so that all carriers would contribute to the costs of providing universal service.

The concept of universal service today is "virtually synonymous with government policies designed to promote the affordability of telephone service and access to the network" (Mueller, 1997). In the United States, targeted subsidies to ensure affordable access to basic telephone service are available for low-income households, regardless of where they are located, as well as for telephone companies providing service in high-cost rural and remote areas, so that the price of service is lowered for all rural households, regardless of income. The Telecommunications Act of 1996 adds two new categories, the first of service, identifying (but not defining) "advanced services" that should be universally available, and second, specifying institutions (namely schools, libraries, and rural health centers) rather than households as the means through which these services should be made accessible (see below). Many other countries have some form of universal service policy, which aims to ensure access to at least basic telephone service throughout the society, including low income, isolated, and disadvantaged residents.[1] In developing countries, such policies are generally formulated as goals to extend access to unserved populations or regions.

ACCESS VERSUS SERVICE

The terms *universal access* and *universal service* are sometimes used interchangeably, although access is becoming the more common term. The distinction is between services that are delivered to all users, and access to a variety of services that may be provided to individual users, households, community facilities, institutions and workplaces, etc. Thus, universal access to the Internet may be accomplished through community telecenters or other public facilities; universal access to telephone services in developing regions can be achieved through installation of public pay phones in neighborhoods and villages.

However, availability of a telephone or a computer connected to the Internet may not suffice for access. The facility must be in a location where the public can use it. A telephone in a chief's house or a computer in a principal's office may not be accessible to all. Facilities must also be affordable; an Internet connection priced beyond the means of most who want to use it or a telephone system with tariffs for major communities of interest (where most people call) that residents cannot afford is not accessible to most of the population they are intended to serve.

Access to the Internet is commonly via public telecommunications networks. In some regions, users have easy access through local "onramps" to information superhighways, whereas in others, they must navigate narrow and congested byways (Parker & Hudson, 1995). As we have seen in the chapter on digital divides, in much of the developing world, there is no access at all. Where local access is available, it is often through a variety of toll roads, many of which charge users based on the distance they travel and time they spend on the networks. The cost of connectivity for many in the developing world is prohibitive. The monthly Internet access charge is 278 percent of average household monthly income in Nepal, 191 percent in Bangladesh, 80 percent in Bhutan, and 60 percent in Sri Lanka, compared to 1.2 percent in the United States. In Africa, the average cost of using a local dial-up Internet account for 20 hours a month is about $68 (including telephone charges but not telephone line rental), while 20 hours of Internet access cost $29 in the United States, $74 in Germany, $65 in Britain, and $52 in France, and all of these OECD countries have per capita incomes at least 10 times greater than the Africa average (OECD, 2003; Mbarika, Jensen, & Miso, 2002).

Access is thus a broader concept than service that involves the following components:

- *Infrastructure:* reach of networks and services, for example, to rural areas, low income populations in inner cities; available bandwidth (such as broadband capacity for high speed Internet access);
- *Range of Services:* for example, basic voice service (plain old telephone service, or "POTS"), value-added services such as ISPs;
- *Affordability:* pricing of installation, monthly service, usage by time or volume, and so on;
- *Reliability:* quality of service, as shown by extent of outages, breakdowns, circuit blockage, circuits degraded by noise or echoes, and so on.

Another important component of access is specification of the entities to whom telecommunications services should be accessible. Users may be considered in several categories:

- *Public:* geographic: urban/rural, regional; demographic: disadvantaged groups such as low income, disabled, ethnic or other minorities;
- *Commercial Enterprises:* large and small businesses, entrepreneurs; critical sectors: such as agriculture, transportation, manufacturing, tourism;
- *Public Services:* health care, education, other government/public services, and so on; nonprofit and nongovernmental organizations (NGOs).

UNIVERSAL ACCESS: A MOVING TARGET

Universal access should be considered a dynamic concept with a set of moving targets. Rapid technological change dictates that the definitions of basic and "advanced" or "enhanced" services will change over time, while the unit of analysis for accessibility may be the household, the municipality, or even institutions such as schools and health centers. Thus, for example a multitiered definition of access could be proposed, identifying requirements within households, within communities and for education and social service providers. For example:

- **Level One:** community access (for example, through shared equipment, kiosks, libraries, post offices, community centers, telecenters)
- **Level Two:** institutional access (schools, hospitals, clinics)
- **Level Three:** household access
- **Level Four:** personal access

Economic and demographic diversity in inner cities, impoverished rural areas, and developing countries will require a variety of goals for information infrastructure. In North America and Europe, the goal has been to provide basic telephone service to every household, with the assumption that businesses and organizations could all afford access to at least this grade of service. However, for Internet access, the United States is applying community and institutional access models. As noted earlier, the U.S. Telecommunications Act of 1996 specifies that "advanced services" should be provided at a discount to schools, libraries, and rural health centers (Telecommunications Act, 1996). Advanced services are currently interpreted as Internet access. In the future, it is likely that advanced services will be redefined, perhaps to include access to new generations of services available through the Internet or its successors. In addition, industrialized countries such as the United States and Canada have extended the concept of basic service beyond quality adequate for voice to include single-party service, and circuits capable of supporting the capacity of current modems, with the assumption that people will want to communicate electronically from their homes. These criteria are also likely to be revised over time to keep pace with demands of the information economy.

Developing countries generally use community access criteria: China, India, Mexico, Nepal, and Thailand, for example, aim for at least one telephone per village or settlement. Other developing countries set targets of public telephones within a radius of a few kilometers in rural areas. The ITU's Maitland Commission called for a telephone within an hour's walk throughout the developing world (International Commission, 1984). Russia's 2004 Law of Telecommunications stipulates a point of access to the Internet in every community of at least 500 people (Gidaspov, 2004).

Personal access is a relatively recent phenomenon, resulting from the proliferation of wireless technologies. Cell phones were originally marketed to business and professional people, but have become the first and only phone for many people in developing regions. Wireless local area networks were designed to create flexible office networks, but the same technology can be used to set up "hot spots" covering parks or villages to enable people with laptop computers or personal digital assistants (PDAs) to access the Internet. Thus "personal access" may also become "public access."

TECHNOLOGICAL TRENDS

In the past, there were few incentives for carriers to provide access to low-income customers such as disadvantaged minorities and inner-city residents who are presumed to have limited demand for new services, and rural and remote regions where the cost of extending or upgrading facilities and services is assumed to be higher than expected revenues. However, technological innovations, many of which were initially designed for other applications, are now creating opportunities to reduce costs and/or increase revenues among these populations. As noted earlier, high capacity wireless may be a less expensive means of providing or upgrading access in urban areas. Wireless also has enormous potential in rural areas where the cost of installing cable or fiber is much higher. Satellite systems may also be used in rural areas for basic telephony (generally through a local network connected to a community satellite terminal) or for Internet access (which could also be through an individual VSAT [very small aperture terminal]). See chapter 7.

Whereas the United States and other industrialized countries must upgrade outdated wireline networks and analog exchanges in rural areas, developing countries can leapfrog old technologies and install fully digital wireless networks. However, availability of technologies that appear appropriate for rural and developing regions does not necessarily mean that they will be installed. Lack of access to ICTs in poor and rural areas of the developing world is often attributed to underinvestment based on assumptions that provision of services in such regions would be "commercially unfeasible" (O Siochru, 1996). Feasibility for the operator requires that revenues exceed costs; thus, greater commercial feasibility can be achieved by increasing revenues and/or reducing costs.

DISCOVERING DEMAND FOR COMMUNICATION SERVICES

In designing networks and projecting revenues, planners often assume that there is little demand for telecommunications in rural areas. Similarly, telecommunications service providers may be reluctant to extend services to poorer populations who are assumed to have insufficient demand to cover the cost of providing the facilities and services. Their forecasts are typically based solely on the lower population densities than are found in urban areas, coupled with a "one size fits all" fallacy that assumes all rural residents are likely to have lower incomes and therefore lower demand for telecommunications than urban residents. Yet, as noted in earlier chapters, rural residents may need telecommunications to order parts and supplies, check on international prices, and arrange transport of their produce to foreign markets. In addition to commercial activities, there may be significant demand from government agencies and NGOs operating in rural areas to administer health care services, schools, other social services, and development projects.

Lack of understanding of demand for telecommunications services (i.e., need for the service and ability to pay) creates problems in designing universal service policy. For example, low-income households in industrialized countries may be able to pay for monthly telephone service subscriptions, which are likely to cost less than other utilities, and less than entertainment, including cable television (Crandall & Waverman, 2000). Thus, it may not be necessary to provide a subsidy for service to most low-income households. However, utilization of the network (for long distance or measured rate calls) which subscribers need to maintain contact for family or work, may be much more expensive. Some consumer representatives have advocated a policy of providing a live telephone connection in all dwellings that could be used for emergencies without charge. Other possibilities include blocks on the line to prevent unauthorized use for other than local calls, and prepaid service in the form of a rechargeable account or smart card that allows calling only up to the limit of funds on the card or in the account. Prepaid cards have become very popular for cellular service in developing regions.

Communications service providers may also be reluctant to extend services to poorer populations who are assumed to have insufficient demand to cover the cost of providing the services and necessary infrastructure. Certainly, household income may be the best indicator of both willingness and ability to pay for communication services. Typically, higher income populations are better educated, and are thus likely to have not only the money but also the skills to use new technologies and services.

Purveyors of new information technologies and services also face the problem of forecasting demand among people who have never used them before. This problem is compounded in developing regions where potential customers may have no familiarity at all with such facilities, and may have very modest or unpredictable incomes. Telecommunications operators therefore tend to base projections on population and per capita income in rural and developing regions. However, operators frequently find that demand in rural and remote areas is greater than forecasts based on population or income alone would indicate.

Revenues from rural telephones may also be greater than expected if incoming as well as outgoing traffic is included. It is important to anticipate the influence of family ties on calling patterns. Communities where many people have left to seek work in the city or overseas may have high volumes of incoming traffic. Mexican laborers call home from the United States; Filipina domestic workers call home from Hong Kong, Singapore, and Malaysia; Indians and Pakistanis call home from the Middle East; and miners in South Africa call their families in other parts of South Africa or neighboring countries such as Mozambique and Zimbabwe. There may also be extensive calling among family members scattered in villages throughout a rural region. For example, the significance of rural toll traffic seems particularly important in estimating rural revenues in India, where revenues from village phones with STD (long distance dialing) are apparently nearly 50 times as high as revenues from phones with only local area access (Telecom Regulatory Authority of India, 2000).

Several approaches can be used to forecast demand in developing regions. Collectively, expenditures on rural and regional telecommunications in developing countries are between 1 and 2 percent of national GDP. One estimate is that rural users in developing countries are able collectively to pay 1 to 1.5 percent of their gross *community* income for telecommunications services (Kayani & Dymond, 1997). The ITU uses an estimate of 5 percent of *household* income as an affordability threshold. To generate reve-

TABLE 6.1
Household Access to Telephones and Television

	Percentage of Households With Telephone	Percentage of Households With TV Set
High income countries	96.1	97.1
Upper middle income countries	59.0	92.6
Lower middle income countries	49.4	89.4
Low income countries	8.2	34.2

Source. ITU, World Telecommunication Development Report, 2003.

nues to cover capital and operating costs, the average household income required would be $2,060; for a more efficiently run network, it would be $1,340. Using the higher estimate, 20 percent of households in low income countries such as Vietnam, Uganda, Kenya, Guinea Bissau, and Ghana could afford a telephone; in lower middle-income countries the range could be from 40 percent to 80 percent, while in upper middle-income countries such as Chile and Eastern European nations, more than 80 percent of households could afford telephone service (ITU, 1998). We are beginning to see evidence of the validity of these estimates through the rapid growth of affordably priced wireless services. (See chapter 7.)

Thus, lack of ICTs cannot necessarily be attributed to lack of demand or purchasing power. Other data may be better indicators of pent-up demand. In many developing countries, television sets are much more prevalent than telephone lines. For example, in industrialized countries, both TV sets and telephone lines are almost universally available. However, in lower middle-income countries there are almost twice as many households with TV sets as with a telephone (including cellular phones), and in low-income countries, there are more than four times as many households with TV sets as with telephones (see Table 6.1). Even in the poorest countries, there may be much more disposable income available than economic data would indicate. It appears that where television broadcasts are available, a significant percentage of families will find the money to buy TV sets. These data may indicate a potential pent-up demand for other communications services.

Other approaches may also be used to gauge demand for information and communication services. For example, the presence of video shops, found even in apparently very poor areas, indicates significant disposable income available for television sets, video cassette players, and cassette rentals. Telephone service resellers (such as in Indonesia, Rwanda, Senegal, and Bangladesh), local cable television operators (common in India), MMDS (Microwave Multipoint Distribution Service) or "wireless cable" found in Thailand and the Philippines, and small satellite dishes on rural homesteads and urban apartments (common in Eastern Europe and many Asian countries) also signal demand and ability to pay for information services.

Cellular operators have recognized this pent-up demand. By making telephone service available in low cost increments using prepaid (or "pay as you go") phone cards, they have outpaced the fixed networks in many developing countries.

RURAL BENCHMARKS NEED NOT BE SET LOWER
THAN URBAN BENCHMARKS

A persistent assumption is that "something is better than nothing" is the only policy technically feasible or economically justifiable for rural areas. However, a corollary of the lessons above, that revenues in rural areas may often be higher and costs lower than assumed, is that it is no longer technically or economically justifiable to set rural benchmarks lower than urban benchmarks for access—both to basic telecommunications and to the Internet.

For example, the U.S. Telecommunications Act of 1996 sets a standard of reasonable comparability: Rural services and prices are to be *reasonably comparable* to those in urban areas. While the United States and other industrialized countries must upgrade outdated wireline networks and analog exchanges in rural areas, developing countries can leapfrog old technologies and install fully digital wireless networks. Thus developing country regulators can also adopt rural comparability standards to avoid penalizing rural services and businesses in access to information services. For example, in the Philippines, after extensive discussion, both government and industry representatives agreed on rural benchmarks including digital switching, single-party service, and line quality sufficient for facsimile and data communications. The industry representatives stated that the new digital networks they were installing in rural areas met those specifications, and that older networks should be brought up to those standards.[2]

STRATEGIES TO EXTEND ACCESS

Service Obligations

Many countries include a universal service obligation (USO) as a condition of the license. The cost of USOs may vary depending on geography and population density. British Telecom's universal service obligation costs just 1 percent of its total revenue base.[3] Some countries such as Chile and Mexico have mandated requirements for operators to install payphones in rural communities; South Africa has also required its wireless operators to install fixed rural payphones. Latin American countries with USOs include Argentina, Chile, Mexico, Peru, and Venezuela. In Mexico, the privatized monopoly operator, TelMex, was to provide service to all communities with at least 500 population by the year 2000, but failed to do so. In the Philippines, local exchange obligations are bundled with cellular and international gateway licenses; licensees were required to install up to 300,000 access lines in previously unserved areas within three years (Hudson, 1997).

Some countries use a "carrier of last resort" model, which has the obligation to provide service if no other carrier has done so. Typically, the dominant carrier bears this obligation and is entitled to a subsidy to provide the service. However, this approach can be flawed if it provides no incentive for the carrier with the USO to use the most appropriate and inexpensive technology and to operate efficiently. It can also serve as a justification for the dominant carrier to be protected from full competition because it has additional costs and obligations not required of new competitors.

Rather than designating a single carrier of last resort, some countries are introducing bidding schemes for rural subsidies. In Chile, a development fund was established in 1994 to increase access for the approximately 10 percent of the population in communities without telephone access. The regulator estimated the required subsidies, distinguishing between commercially viable and commercially unviable, and put them out to competitive tender. There were 62 bids for 42 of the 46 projects. Surprisingly, 16 projects were awarded to bids of zero subsidy; as a result of preparing for the bidding process, operators were able to document demand and willingness to pay in many communities. Once completed, these projects were to provide service to about 460,000 people, about one third of the Chilean population without access (ITU, 1998). Peru is introducing a similar program.

Targeted Subsidies

A variety of schemes can be used to subsidize carriers that serve regions where revenues would apparently not cover costs. Subsidies may be paired with USOs to compensate the carrier with the obligation to serve. The traditional means of ensuring provision of service to unprofitable areas or customers has been through cross-subsidies, such as from international or interexchange to local services. However, technological changes and the liberalization of the telecommunications sector now make it impracticable to rely on internal cross-subsidies. For example, customers may bypass high priced services using so-called "callback"[4] services or Internet telephony.

In a competitive environment, cross-subsidies cannot be maintained. Carriers that have relied on revenues from one service to subsidize another now face competitors that can underprice them on individual services. Also, new entrants cannot survive if their competitors are subsidized. Therefore, if subsidies are required, they must be made explicit and targeted at specific classes of customers or locations such as:

• **High cost areas:** Carriers may be subsidized to serve locations that are isolated and/or have very low population density so that they are significantly more expensive to serve than other locations. This approach is used in the United States and Canada.

• **Disadvantaged areas or customers:** Subsidies may target economically disadvantaged areas or groups that could not afford typical prices for installation and usage, or where demand for service is significantly lower than average. Some carriers may offer interest-free loans or extended payment periods to assist new subscribers to connect to the network. In the United States, the Lifeline program subsidizes basic monthly services charges for low-income subscribers. The subsidy funds come from a combination of carrier contributions and surcharges on subscriber bills. Some 4.4 million households receive Lifeline assistance. Also in the United States, the Linkup program subsidizes connection to the network for low-income households.

• **Route Averaging:** Some countries including Australia, Canada, the United Kingdom, and the United States require that rates be averaged so that all customers pay uniform distance charges, regardless of location. Thus, for example, the rate per minute between Sydney and Melbourne would be the same as the rate over an equal distance in the Australian Outback, where costs are much higher. Such policies can bridge the digital divide by reducing rural access costs.

FUNDING SUBSIDIES

Funds for subsidies may be generated from several sources including contributions required from all carriers such as a percentage of revenues, a tax on revenues, or a surcharge on customer bills. Subsidies may also come from general tax revenues or other government sources.

• **Transfers Among Carriers:** Some countries with many carriers rely on settlement and repayment pooling schemes among operators to transfer payments to carriers with high operating costs. For example, the U.S. Universal Service Fund is mandated by the Federal Communications Commission (FCC) but administered by the carriers through the National Exchange Carriers Association (NECA), and transfers funds to subsidize access lines to carriers whose costs are above 115 percent of the national average.[5]

• **Government-Financed Funds:** In Poland, more than 7,885 localities were connected between 1992 and 1996 with funding of US$20 million from the state budget (ITU, 1998). In 1994, Peru established a rural telecommunications investment fund, FITEL (Fondo de Inversion de Tele-comunicaciones), which is financed by a one percent tax on revenues of all telecommunications providers, ranging from the country's newly privatized monopoly operator, Telefonica/Entel, to cable TV operators. Since established, it has generated an average of US$450,000 per month, growing by US$12 million annually (ITU, 1998). Private sector operators may apply to FITEL for financing (Kayani & Dymond, 1997, pp. 63–64).

COMMUNITY ACCESS: TELECENTERS

An increasingly popular means of providing community access is the telecenter, a term that has been used to refer to a variety of means of providing access to ICTs, ranging from cybercafés to facilities located in public buildings such as libraries and post offices, to standalone public access centers. Telecenters originated in Scandinavia in the 1980s as telecottages, with the goal of helping to diversify rural economies through enabling rural residents to become information workers, by taking on projects such as word processing, data entry, telemarketing, and interviewing from their communities for urban clients (see chapter 5). With the advent of public access to the Internet, telecenters evolved into a means for community residents to use shared facilities to send e-mail and access the Web. The idea spread to developing countries where access to computers and telecommunications is often very limited and costly, especially in rural areas. In industrialized countries, telecenters have been seen as a means of extending Internet access to rural and isolated residents, poor inner-city residents, indigenous peoples, migrants, and other disadvantaged groups.

Initiatives to support public Internet access through community telecenters have been supported by many development agencies such as the ITU, UNDP, Canada's International Development Research Centre (IDRC), the U.S. Agency for International Development (USAID), and the World Bank. Governments may provide support for community access through universal service funds, special "digital divide" initiatives, or other programs designed to extend Internet access to unserved populations. In general, the projects are designed to develop sustainable models to meet the information and communication needs of the communities, with the assumption that these models will likely evolve during and following implementation. Many target specific groups such as youths, women, minorities, and senior citizens.

Most telecenters are established in existing buildings such as shops, cafes, community centers, schools, or libraries. However, some are designed to be self-contained units. For example, LINCOS (Little Intelligent Communities), an initiative of the Costa Rica Foundation for Sustainable Development, are self-contained digital community centers, with computers and communications equipment as well as optional facilities for telemedicine and videoconferencing installed in a modified shipping container. LINCOS projects are adapted to each community, and include training and follow-up to foster developmental applications and sustainability.[6]

South Africa has funded the installation of stand-alone telecenters equipped with phone lines, facsimile, and computers with Internet access through a Universal Service Fund; South Africa now plans to provide Internet access to government information and electronic commerce services through post offices. International development agencies have funded community telecenters in many other African countries, such as Benin, Ghana, Mali, Mozambique, and Uganda.

In Indonesia, the government has provided public telephone access through community access points called WarTels. Many of these have evolved into WarNets that also provide Internet access. Clients include Open University students who can use the WarNets to submit assignments and communicate with instructors.

In Peru, the Red Cientifica Peruana (Peruvian Scientific Network) has set up a national network of 27 telecenters, each of which typically consists of 20 desktop computers with dedicated Internet access. The telecenters provide computer rentals, training, personal e-mail accounts, World Wide Web page development, and other services. RCP is completely self-sustaining, receiving no subsidies from government or other sources.[7] Although it started as a network for the exchange of academic and scientific information, RCP developed a successful business model, with a fee structure that covers costs and allows its organizations to grow while retaining profits.[8]

In the United States, Community Technology Centers' Network (CTCNet) consists of more than 1,000 community technology centers where people in low-income communities can access computers and computer-related technology, such as the Internet. For more than a decade, CTCNet's mission has been to support community technology centers so that they can better serve their constituencies. Start-up support has come from government grants and corporate donations.[9]

Sustainability, or simply survival after the pilot phase, has been a struggle for many telecenters. Some sell other services such as photocopying and desktop publishing; some become outlets for e-government services or contract to provide computer training. A few have become ISPs, offering connectivity for others in their community. Notably, the Red Cientifica Peruana has become the largest ISP in Peru and has developed the country's most popular portal with 20,000 hits per day. RCP also plans to expand into long distance telephony and content production.[10] The telecenter in

Timbuktu, Mali, also became an ISP for small businesses and NGOs that needed connections to the Internet.

INTERNET ACCESS THROUGH SCHOOLS

Schools can also be important means of extending Internet access. In developing regions, organizations such as Schoolnet Africa, the World Bank's World Links for Development program and the Commonwealth of Learning (COL) provide facilities and training for schools, and support the development of indigenous content. Many countries have adopted policies to connect all of their schools to the Internet, but most developing countries are struggling to do so. Organizations such as Schoolnets can help to obtain funds and equipment, and influence connectivity policies. In some industrialized countries, government funding covers the ongoing costs of Internet connections for schools.

An approach used in the United States that could be a model for developing countries is the E-Rate[11] (short for "education rate") program that provides discounts for Internet connections to schools and libraries, with revenues from the Universal Service Fund. (For more details, see the boxed text on p. 98.) Despite start-up administrative problems and later lack of sufficient oversight, the E-Rate program has been largely successful in facilitating access for schools and libraries. In addition, it has provided new customers for telecommunications operators. In some cases, the schools and libraries have attracted Internet service to some previously unserved rural communities or inner cities because they act as "anchor tenants" (major ongoing customers) for service providers.

Other countries with universal service policies generally subsidize the carrier directly to install facilities or provide services at a reduced price. For example, the "carrier of last resort model" requires that an incumbent provide service in areas that are considered uneconomical to serve. A new or competing carrier may be required to provide service in a rural area as part of its license obligation. Carriers may also be required to provide a discount for Internet access to schools and/or other designated entities as a condition of their license. These approaches create no incentives for new entrants or for keeping construction and maintenance costs down. A carrier that can negotiate a high subsidy will have no incentive to reduce equipment or operating costs. Quality of service may also suffer if the facilities and services are perceived to be of low value, or the carrier concludes there is little likelihood of sufficient demand without the subsidy.

In contrast, the E-Rate model provides the full approved tariff rate to the carrier providing the service (part of the fee coming from the USF and the rest from local sources), rather than requiring the carrier to lose money on the service. It provides incentives for pricing efficiency by requiring the school to post its requirements on the Universal Service website in order to solicit competitive bids for qualifying facilities and services.

Telecenters: Bulgaria's PC3 Project

Between 2000 and 2002, the United States Agency for International Development (USAID) supported the establishment of telecenters known as PC3s (for Public Computer and Communications Centers) in 10 small towns in Bulgaria. Innovative strategies in implementation included:

- Selection of two people with entrepreneurial backgrounds as managers of each PC3 (several management teams were married couples);
- A training workshop that included business planning as well as technical and administrative skills;
- Involvement of telecenter operators in customizing support and undertaking part of the start-up costs;
- The provision of prepaid vouchers to be distributed among target users to promote PC3 usage and develop a client base.

The PC3 project had goals of both increasing access to information through ICTs for people in these small towns and being sustainable. To promote sustainability, the project recruited managers who had some entrepreneurial experience, provided training in business planning, and let them develop their own business plans. Yet telecenters based only on a commercial model might exclude the target populations that USAID was trying to reach: youths, disadvantaged populations, teachers, social service workers, senior citizens, etc., who might not have the discretionary income or the skills or confidence to utilize the telecenter. To reach these target groups, the project provided each telecenter with a budget of prepaid vouchers or coupons so that operators could offer a limited number of free introductory services to desirable groups for "common good" purposes. Used coupons were reimbursed from project funds, thus also providing a source of income during the pilot phase (Tifft, 2002).

The evaluation after the two-year start-up period found that "The pilot project has successfully demonstrated that the multipurpose telecenter can be a potentially sustainable and replicable demand-driven small private enterprise model for human capacity development in a transitional economy country . . . ." (Tifft, 2002, p. 4). The telecenters were becoming sustainable businesses, although they had not spun off microbusinesses as had been anticipated. Some had branched into other ICT activities; half of them became ISPs and built municipal wireless LANs, thus increasing Internet penetration in Bulgaria. Operators also learned to adapt services to meet local needs, including ICT training.

Targeted Subsidies: The E-Rate[12]

The U.S. Telecommunications Act of 1996 expanded the original purpose of the Universal Service Fund, established to extend reasonably priced telephone services to rural and other underserved areas, to include support for telecommunications services for schools, libraries, and rural health care providers. The E-Rate[13] (short for "education rate") is the program that provides discounts on a wide variety of telecommunications, Internet access and internal connections products and services for schools and libraries. The Federal Communications Commission sets the overall policy for the program, which is administered by a nonprofit entity, the Universal Services Administrative Company (USAC). Funds come from telecommunications carriers, which are required to contribute a set portion of their revenues to the Universal Service Fund (USF). Carriers generally pass through these costs to customers through itemized charges on their telephone bills. The FCC makes payments from this central fund to support the Schools and Libraries program, as well as three other Universal Service programs: Low-Income (customers), High-Cost (service areas), and Rural Health Care.

Up to $2.25 billion worth of discounts can be made available each year. Schools may apply for all "commercially available telecommunications services" ranging from basic telephone services to T-1 and wireless connections, Internet access including e-mail services, and internal networking equipment. Discounts are not available for computers (except network servers), teacher training, and most software.[14] The E-Rate subsidy is targeted not just to the school, but to the students because it includes funding for internal networking as well as connection to the Internet. The goal, according to former FCC Chairman Reed Hundt, was to ensure that the Internet got to the classroom, and not just to the schoolhouse door (Hundt, 2000). Approved costs are billed directly to USAC, up to the limit of the subsidy. Schools and libraries are responsible for the remainder, and must demonstrate that they can cover their portion of the costs. Discounts range from 20 percent to 90 percent, with the largest discounts for schools in low-income rural and urban areas.

Schools must prepare a technology plan that must be approved by the state before they are eligible to apply for E-Rate funds. The purpose of this requirement is to ensure that school staff consider issues such as sources of funding for other equipment and maintenance, training for teachers and students, and strategies for integrating use of computers and the Internet into the curriculum. The intent is to attempt to assure that the facilities will be constructively utilized and the facilities will be sustainable (i.e., schools will commit to finding funds and people to cover additional costs of operations, maintenance, and applications). Once the school's application is approved, its requirements are posted on the Universal Service Corporation's Web site (www.universalservice.org) for 28 days, following which it may select from competitive bids or negotiate with the carrier serving the area according to E-Rate procurement rules and guidelines.

An important element of the E-Rate program is that it is designed to be incentive-based. Subsidies are not awarded directly to the carrier but through the user—it is the school or library which is eligible to receive the discount. Like a voucher system, the E-Rate can empower schools and libraries because they now have resources for technology, rather than being relegated to the sidelines by the telecommunications industry as unlikely or undesirable customers.

NOTES

1. For a comparison of universal service in the United States, Canada, and the United Kingdom, see Crandall and Waverman (2000).
2. Meeting at Department of Transport and Communications attended by the author, Manila, January 1998.
3. OFTEL (Office of Telecommunications). *A Framework for Effective Competition.* London: OFTEL (1994), quoted in Kayani and Dymond (1997, p. 53).
4. The simplest form of callback involves a caller where tariffs are high asking the other party to call back using cheaper rates from their location. Simple examples include a traveler using a hotel's payphone to call home and asking family members to call back to her room, rather than paying the hotel's high charges. Now callers in high-cost areas (typically in developing countries) can set up an account with a "callback" company and then call a special number, hang up, and wait for the return call from the cheaper jurisdiction. The company can identify the caller without even answering the call, and can set up the call between the two parties using its own software.
5. See www.neca.org and information on the Universal Service Fund on the FCC's Web site, www.fcc.gov
6. See www.lincos.net
7. See http://ekeko.rcp.net.pe/
8. Harvard University Center for International Development. See http://cyber.law.harvard.edu/readinessguide/vignettes.html
9. See www.ctcnet.org
10. See http://ekeko.rcp.net.pe/
11. Also known as the Snowe-Rockefeller-Exon-Kerrey Amendment.
12. See Hudson (2004) for a more detailed analysis of the E-Rate.
13. See endnote 11.
14. For details, see www.sl.universalservice.org

Chapter 7

Technologies for Extending Connectivity

Within four years virtually every African household will either have cellular access or live next door to someone who does.

—Carl Edgar (2001)

TECHNOLOGICAL TRENDS

As noted in chapter 1, several key technological trends are driving the proliferation of new information and telecommunications services. They include: capacity (increase in bandwidth); digitization (so that any service is transmitted as a stream of bits); ubiquity (potential availability of services anywhere, primarily by terrestrial wireless and satellite technologies); and convergence (combining voice, data, and images delivered over a single network). Many new technologies offer the possibility for developing countries to leapfrog earlier generations of equipment to close the information gap. This chapter examines technologies that are being implemented in developing regions, or could be adapted for rural and developing country applications. It then examines trends in connectivity costs, and steps in planning, training and technical support that could further reduce costs.

THE BACKBONE AND THE LAST MILE

Increasing access to telecommunications services ranging from voice communications to the Internet requires investment both in backbone networks and in facilities to reach end users; this latter component is often called the "last mile." Backbone connectivity remains a problem in some developing regions; for example, some countries in the Middle East and Africa have very limited links with the outside world. However, in many other developing regions, particularly where competition is allowed in international gateways, investment has greatly increased capacity.

While there may now be capacity gluts in transcontinental optical fiber capacity, there can be shortages in capacity linking communities of interest (countries that share common cultural, linguistic, or economic ties). For example, "a computer user in Cairo who wants to listen to a sermon via streaming audio from a Web site in Qatar must use a connection that runs from Cairo to Amsterdam to Atlanta then back across the Atlantic into the Persian Gulf. The result is a painfully slow Internet connection" (Romero, 2001). However, connecting end users remains a much bigger problem. Even in many industrialized countries, most Internet users rely on slow dial-up connections. In the developing world, where telephone connections themselves are scarce, the problem is often providing any connectivity.

THE WIRELESS CONNECTION

The gap in access to basic communications (for voice, fax, and short text messages or e-mail) is beginning to close. It took a century to connect the first billion people by wire, but only a decade to connect the second billion, thanks largely to the growth of wireless networks. Yet a major disadvantage of most current wireless systems is very limited bandwidth that makes them impractical for accessing the World Wide Web. New cellular technologies (often referred to as Third Generation or 3G) promise more bandwidth, but installation costs are high.

In developing countries without sufficient infrastructure, wireless technology can be used to provide primary service, substituting for traditional wireline telephone service. In many developing regions, wireless growth has been explosive, with mobile phones becoming the first telephones for most new subscribers. In all developing regions, there are now more wireless subscribers than fixed lines. See Table 7.1.

Wireless technologies have the potential to increase connectivity much faster than wireline because it is not necessary to install wire to each subscriber. A single wireless antenna may cover a neighborhood, a town, or, if terrain is suitable, a whole district or county. Wireless installations can be used to provide public access. For example, new cellular operators in South Africa were required to install 30,000 wireless payphones within five years as a condition of the license (ITU, 1998, p. 53). Alternatively, a wireless subscriber may resell access. Entrepreneurs in Bangladesh offer payphone service using cell phones leased from Grameen Phone, which they carry by bicycle to various neighborhoods.

TABLE 7.1
Developing Countries/Emerging Economies
With More Wireless Than Fixed Lines

Region	Number of Countries	Percentage of Countries in the Region
Africa	40	78.4
Asia	28	62.2
Latin America/Caribbean	24	72.7
Eastern Europe	13	72.2

Source. Derived from ITU data for 2003.

The proliferation of information resources on the Internet's World Wide Web has driven demand for higher bandwidth—for faster access to Web sites, viewing graphics, transmission of large attachments to e-mail messages, downloading audio and video files, and so forth. Demand in industrialized countries is primarily for connections from laptop computers and handheld devices (for example, personal digital assistants [PDAs] in North America and mobile phones in Asia and Europe). Broadband wireless technologies designed for these applications may also be the most cost-effective solution for developing regions where there are no alternatives such as DSL or cable.

TERRESTRIAL WIRELESS TECHNOLOGIES

• **Cellular:** Cellular telephony has become the first and only telephone service for people in many developing countries, where it may be available much sooner than fixed line service. In countries such as Côte d'Ivoire, Gabon, Rwanda, Tanzania, Uganda, Cambodia, Mongolia, Indonesia, and the Philippines, there are now more cellular telephones than fixed lines. Where there are no fixed lines, a cell phone with a cellular modem can be used to allow access to the Internet. For example, the community tele-center in Buwama, Uganda, about 60 km from Kampala, connects to the Internet via cellular modem. However, cellular access is often quite costly, and bandwidth is limited, so mobile phones are likely to be more practical for short bursts of traffic for e-mail and text messages than for surfing the Web.[1]

General Packet Radio Service (GPRS): GPRS is a digital mobile technology, also referred to as 2.5G technology. This technology allows mobile phones to be used for sending and receiving data over an Internet Protocol (IP)-based network. GPRS-enabled networks can run up to 115 kbps, compared with GSM speeds of only 9.6 kbps. GPRS offers "always-on," higher capacity, Internet-based content and packet-based data services.[2]

3G Mobile Services: Third generation (3G) mobile networks are beginning to be introduced in some industrialized countries, but generally not in rural or isolated areas. They offer greatly increased bandwidth over existing mobile networks, with the possibility of Internet access to handheld devices such as portable phones, personal digital assistants, and small personal computers. However, the capital cost of upgrading existing networks to 3G is very high, and the price of access for Internet applications may be greater than for other options. In some countries, existing networks are being upgraded to 2.5G, which has considerably more bandwidth than earlier generations, but such upgrades have not generally occurred in rural areas.

- **Wireless Local Loop (WLL):** Wireless local loop systems can be used to extend local telephone services without laying cable or stringing copper wire. For example, instead of having a fixed line connection, schools would have a wireless link to the telecommunications network. WLL costs have declined, making it competitive with copper. Wireless allows faster rollout to customers than extending wire or cable; it also has a lower ratio of fixed to incremental costs than copper, making it easy to add more customers and to serve transient populations. Wireless is also less vulnerable than copper wire or cable to accidental damage or vandalism. Examples of developing countries and emerging economies with WLL projects include Bolivia, Czech Republic, Hungary, Indonesia, South Africa, and Sri Lanka (ITU, 1998, p. 53).

- **WiFi:** It is now possible to set up inexpensive wireless local area networks using a technology called WiFi (for wireless fidelity) or 802.11 (an IEEE standard). Most laptop computers now come equipped with WiFi ports, and inexpensive adapter cards are also available. WiFi "hotspots" in industrialized countries provide high-speed access for hotels, businesses, college campuses, and cybercafés. A WiFi hotspot could cover an entire village. Enthusiasts in the United States and some other industrialized countries are working on projects to daisy-chain these nodes to provide free or very low cost connectivity over a wide area. This approach may be more attractive in terms of cost and fast deployment than 3G networks in developing regions.

- **WiMAX:** WiMAX (IEEE 802.16) technology can be used for connecting WiFi hotspots and wireless LANs to the Internet, providing campus or village connectivity, and providing wireless last-mile broadband access to the Internet as an alternative to DSL and cable modems. WiMAX can cover up to 50 kilometers, and does not require direct line of sight with the base station.[3] In developing regions, WiMAX could serve as a last-mile wireless local loop solution or a link between a village or neighborhood WiFi hotspot and the backbone network.

- **CDMA450:** Operating in the 450 MHz band, CDMA 450 utilizes a larger cell size compared to services in other bands, translating into fewer cell sites and lower capital and operating expenditures. Its propagation characteristics and resulting greater coverage area may make it an appropriate wireless local loop solution for rural and developing regions.[4] It also uses different frequencies from those assigned to GSM, so that it could be added as a broadband access technology in regions already using low-bandwidth GSM.

- **Point-to-Point Wireless Systems:** If the telephone company does not provide WLL, schools and other organizations and businesses may be able to install or lease their own wireless links to the Internet. Point-to-point fixed wireless such as microwave systems can provide high-speed Internet access by connecting to an ISP's point of presence (POP). These fixed wireless links may be the least expensive means of getting high-speed Internet access if wireline services are not available. Time division multiple access (TDMA) radio systems are a means of providing wireless rural telephony. They typically have 30 to 60 trunks and can accommodate 500 to 1,000 subscribers. Their range can be extended using multiple repeaters. Such systems are found in some rural areas in Canada and the United States and in some developing countries (Kayani & Dymond, 1997).

- **Cordless:** Short range cordless extensions can provide the link from wireless outstations to subscriber premises; the DECT (Digital European Cordless Telephone) technology standard can also allow the base station to act as a wireless PBX and further reduce cost (Kayani & Dymond, 1997, p. 48). For example, DECT has been used in South Africa to provide links to rural pay telephones and telecenters. However, DECT has very limited bandwidth, so that it is not suitable for accessing the World Wide Web.

- **WAP (Wireless Access Protocol):** This protocol has been developed to make it possible to transmit Web pages and other data to cellular phones. GPRS-enabled phones may use WAP. It may eventually be adapted for wireless services in developing regions so that Internet information can be transmitted to low-bandwidth wireless systems. However, the variety of Web content accessible through WAP-enabled devices is still very limited.[5]

SATELLITE TECHNOLOGIES

Satellite facilities can also be installed where communications is needed, particularly in remote and isolated areas, rather than waiting for terrestrial networks to be extended from the cities. While satellites are used in industrialized countries primarily for broadcasting reception, they can also provide telephony in isolated regions where distances are too great for terrestrial wireless. Geostationary satellites (GEOs) may provide radio and television reception as well as interactive voice and data, although the interactive link has a quarter second delay in each path (to and from the satellite) because the satellite is located 36,000 km (or 22,300 miles) above the earth. Low earth orbiting (LEO) satellites have no noticeable delay because they are located only about 800 miles above the earth. However, current generations of LEOs have very limited bandwidth, making them suitable only for voice and text or simple e-mail. Satellite facilities particularly relevant for rural and developing regions include:

• **Very Small Aperture Terminals (VSATs):** Small satellite earth stations operating with geosynchronous (GEO) satellites can be used for interactive voice and data, as well as for broadcast reception. VSATs have been used for interactive services such as videoconferencing and for broadcasting in the Australian Outback. VSATs are also used for telephony and Internet access in Alaska. Banks in remote areas of Brazil are linked via VSATs; the National Stock Exchange in India links brokers with rooftop VSATs. VSATs for television reception (known as TVROs for television receive only) deliver broadcasting signals to viewers in many developing regions, particularly in Asia and Latin America.

• **Demand Assignment Multiple Access (DAMA):** In GEO satellite systems, instead of assigning dedicated circuits to each location, DAMA allows the terminal to access the satellite only on demand and eliminates double hops between rural locations served by the same system. The system is very cost effective because satellite transponder expense is reduced to a fraction of that associated with a fixed-assignment system for the same amount of traffic. Also, digital DAMA systems provide higher bandwidth capabilities at much lower cost than analog. DAMA is used to reduce costs of village telephone service in Alaska, and is now used in other satellite dial-up networks.

• **Internet via Satellite:** Internet gateways can be accessed via geostationary satellites. For example, Alaskan villagers and residents of the Canadian Arctic are connected to the Internet via U.S. and Canadian domestic satellites. A Mongolian ISP called MagicNet and some African ISPs access the Internet in the United States via PanAmSat, a global satellite system that competes with Intelsat. However, these systems are not optimized for Internet use, and may therefore be quite expensive. Also, as noted earlier, there is a half-second delay in transmission via GEO, although it is a more obvious hindrance for voice than data. Several improvements in Internet access via GEOs are becoming available:

Hybrid Systems: A system designed by Hughes known as DirecPC (now called DIRECWAY) uses a satellite to deliver high bandwidth Internet content downstream to a VSAT from an ISP. Upstream connectivity is provided over existing phone lines. This approach is designed for rural areas where there is telephone service, but bandwidth is very limited. Some rural schools in the United States use DIRECWAY for Internet access.[6]

Interactive Access via VSAT: Several companies are developing protocols for fully interactive Internet access via satellite, to make more efficient use of bandwidth and thus lower transmission costs for users. Examples include Gilat, VITACom, Tachyon, and Aloha Networks.[7] Hughes and WildBlue now offer a two-way satellite service in the United States, and direct satellite Internet access is available on domestic and regional satellites in other regions.

Bandwidth on Demand: Constellations of LEO (low earth orbiting) satellites such as McCaw's Teledesic (development of which has been suspended) and Alcatel's Skybridge, and new generations of GEOs such as Loral's Cyberstar and Hughes' Spaceway are designed to offer bandwidth on demand for Internet access and other broadband services.

- **Satellite Telephony:** Global Mobile Personal Communications Systems (GMPCS) operated by Iridium and Globalstar use LEO satellites to provide telephone service almost anywhere in the world to handheld receivers. These systems provide voice and low-speed (typically 2.4 to 9.6 kbps) data connectivity that could be used for e-mail but is too slow for Web access. The price per minute for these services is typically much higher than national terrestrial services. Thus while they may be used by reporters and adventurers, and by relief agencies to coordinate aid and logistics for natural disasters, they are far too expensive for developing country residents. A somewhat less expensive option for the Middle East and much of Asia and Africa is phone service using Thuraya, a GEO satellite with operations headquartered in Abu Dhabi.

- **Data Broadcasting by Satellite:** GEO satellites designed for interactive voice and data can also be used for data broadcasting. For example, China's Xinhua News Agency transmits broadcasting news feeds to subscribers equipped with VSATs. The WorldSpace satellite system delivers digital audio directly to small radios. Although one market for these products is people who can afford to subscribe to digital music channels, the system can also be used to transmit educational programs in a variety of languages for individual reception or community redistribution. It can also be used for delivery of Internet content; schools or telecenters can identify which Web sites they want to view on a regular basis, and WorldSpace broadcasts the data for reception via an addressable modem attached to the radio. World-

Space has donated equipment and satellite time for pilot projects at schools and telecenters in Africa.[9]

• **Store-and-Forward Messaging:** For developing regions, Volunteers in Technical Assistance (VITA) developed a satellite-based system called VITAsat, capable of delivering low-cost communications and information services to remote communities. The system uses simple, reliable, store-and-forward e-mail messages relayed to the Internet via LEO satellites. Using compression technology and software that allows access to Web pages using e-mail, VITAsat can make the Internet accessible virtually anywhere. VITA's two satellite system has the capacity to serve about 2,500 remote rural terminals that could be installed in schools, clinics, community centers, and NGOs.[10]

WIRELINE TECHNOLOGIES

Innovations in wireline technology make it possible to provide high-speed Internet access over telephone lines, rather than having to upgrade existing copper networks. These technologies are typically available only in cities, large towns, and high-density rural areas.

• **Digital Subscriber Line (DSL):** Several variations of DSL technology have been developed that provide data rates of 384 kbps to 1.544 mbps downstream over existing copper pair for services such as limited video-on-demand and high-speed Internet access. This technology is replacing ISDN in industrialized countries because of its greater bandwidth. It can be used in urban areas where copper wire is already installed, but its range is limited to about 1 km from a telephone exchange. However, copper wire is prone to theft in some countries; Telkom South Africa reported more than 4,000 incidents of cable theft in one year, at an estimated cost of about US$50 million (ITU, 1998).

• **Integrated Services Digital Network (ISDN):** Regular twisted pair copper telephone lines can carry two 64 kbps channels plus one 16 kbps-signaling channel. One channel can be used for voice and one for fax or Internet access, etc.; or two can be combined for videoconferencing or higher speed Internet access. ISDN was developed in Europe, and may be available from telephone companies in some urban and suburban areas of developing countries.

• **Cable Modems:** Some cable television systems can also be used for high-speed Internet access via cable modems. Like DSL, cable offers much higher bandwidth than dial-up telephone lines. However, a high volume of users may result in congestion of a shared cable network, and older networks may not be easily converted for two-way connectivity.

• **Hybrid Fiber/Coax (HFC):** A combination of optical fiber and coaxial cable can provide broadband services such as TV and high-speed Internet access as well as telephony; this combination is cheaper than installing fiber all the way to the customer premises. Unlike most older TV cable systems, HFC allows two-way communication. The fiber runs from a central telephone switch to a neighborhood node; coaxial cable links the node to the end user such as a school. Developing countries with HFC projects include Chile, China, India, and Malaysia (ITU, 1998, p. 57).

• **Optical Fiber:** Optical fiber is commonly found in links between switches, terrestrial backbone networks, and submarine cables. The advantage of fiber is its enormous bandwidth, which can be used for high capacity trunking but also for high-speed Internet access or other services such as videoconferencing. Fiber is being installed in new towns and subdivisions in industrialized countries, but the cost of upgrading existing local loops to fiber is very high. However, Singapore has provided fiber to end users, and Japan is committed to upgrading local loops to optical fiber throughout the country. Telephone companies upgrading their networks may install optical fiber to institutional customers such as hospitals, schools, and businesses, but the price of access may be prohibitive. Some schools have managed to gain free or heavily discounted access to so-called "dark fiber," excess capacity which has been installed but is not in use.

ELECTRONIC MESSAGING

• **Voice Messaging:** Voice mail systems can be used to provide "virtual telephone service" to people who are still without individual telephone service. Callers can leave messages in rented voice mail boxes, which the subscribers can retrieve from a pay phone. For example, TeleBahia in northeastern Brazil used voice messaging technology to offer virtual telephone service to customers still without individual telephone lines. Small businesses could rent a voice mail box for a monthly fee and check their messages from a payphone, providing a means for clients to contact them. Similarly, in Africa an investor set up wireless public payphones and provided voice mail accounts and pagers that announce incoming messages. The recipient would be able to return the call or leave a voicemail message using a phone card. The same approach has been used in California for migrant farm workers to enable them to stay in touch with their families, and in homeless shelters to enable job seekers to be contacted by employers.

• **Electronic Mail:** E-mail is much faster than the postal service and cheaper than facsimile transmission. E-mail has become an important business and organizational tool in developing regions. Those without telephones or computers in their workplace or home are increasingly using telecenters and cybercafés to exchange e-mail. Some families in South Asia keep in touch with relatives who are migrant workers in the Middle East by e-mail, as it is less expensive than international phone calls. Of course literacy is required, but generally a literate family member or friend can be found to help send and read messages.

• **Text Messaging:** Text messages on cellphones (known as Short Message Service or SMS) can substitute for many e-mail functions. Text messages are often less expensive than voice calls, making them very popular in many developing countries. For example, the Philippines is now the world's largest user of SMS.

OTHER COMMUNICATION TECHNOLOGIES

Other technological innovations that can be used to improve access to communication networks in rural and developing regions include:

• **Internet Telephony (Voice Over IP):** Packetized voice can be transmitted very inexpensively over the Internet. The advantage of using Internet protocols for voice as well as data is much lower transmission cost than over circuit-switched telephony networks. Some carriers are now offering dial-up access to Internet telephony. For example, China has built parallel Internet protocol (IP) networks that individuals can access from existing payphones using prepaid phone cards. Some telecenters in developing countries are apparently offering voice over IP (Internet protocol) services; however, in many developing countries voice over IP is considered illegal.

• **Digital Compression:** Compression algorithms can be used to "compress" digital voice signals, so that eight or more conversations can be carried on a single 64 kbps voice channel, thus reducing transmission costs. Compressed digital video can be used to transmit motion video over as few as two telephone lines (128 kbps), offering the possibility of low-cost videoconferencing for distance education and training, and other interactive functions such as counseling, family contacts, and administrative meetings.

• **Rural Exchanges:** Advances in microprocessing are reducing switching costs, even for small rural installations. Rural exchanges can be stand alone or remote switching units (RSUs) from a larger exchange. Modified PBXs (private branch exchanges) can also be used to minimize costs. Concentrators reduce transmission costs by multiplexing calls for transmission to the exchange. India's C-DOT (Centre for Development of Telematics) has pioneered in producing such equipment for rural India and other developing countries.

• **Smart Cards:** Prepaid phone cards, widely available in Europe and Asia, have been introduced in developing countries to eliminate the need for coin phones (which require coin collection and may be subject to pilferage and vandalism). Cellular operators have extended this concept to offer prepaid cellular service using rechargeable smart cards, so that telephone service is now available to customers without credit histories or even bank accounts. In South Africa, Vodacom sold more than 300,000 prepaid "pay as you go" starter packs and one million recharge vouchers for cellular use in a year (ITU, 1998, p. 44). In Uganda, within one year of licensing a second cellular operator, aggressive marketing of prepaid service and attractive pricing resulted in there being more cellular customers than fixed lines in the country. Stored payment cards could potentially be used for other services such as Internet access.

• **Community Radio:** While not a new technology, small community radio broadcasting stations can be important news sources for the community and can be used to broadcast educational radio programs for listening both in school and at home or community centers (e.g., Latchem & Walker, 2001). Some school and telecenter projects are combining computer facilities with community radio stations, so that information received via the Internet can be communicated more widely. Portable wind-up or solar-powered radio receivers are practical for school and community use.[11]

TRENDS IN CONNECTIVITY COSTS

Key to determining financial viability and pricing of rural telecommunications are the costs of installing, operating, and maintaining the facilities. Technological innovations such as those identified earlier are helping to reduce costs per line; however, costs range widely because they are very sensitive to subscriber distribution and density. Costs may range from below US$1,000 per line using wireless loop technology where subscriber densities exceed 0.5 per square kilometer within 40 kilometers (25 miles) of an urban center, to US$2,000 to US$3,000 per line over a more widespread area using wireless loop and multiaccess subscriber radio (Kayani & Dymond, 1997, p. xv). TDMA multiaccess radio, cellular/wireless and rural wireline exchanges, can provide rural telecommunications service for about US$1,500 per line under optimum conditions, dropping to $1,000 per line under ideal conditions (p. 40).

Costs can be higher in very thinly populated areas, and when using satellite technology. However, as noted above, technological advances are also reducing satellite costs. In countries with very low population density, such as parts of Bolivia, Peru, the Philippines, and countries in the African Sahel region, costs can exceed US$20,000 per line. Countries with high rural population density such as India, Bangladesh and parts of Indonesia have much lower capital costs, and therefore can support higher teledensity despite very low rural per capita incomes.

Operating costs must also be taken into account. The ITU (1998) estimates annual operating costs at $200 to $750 per line; with a median value $300; and best practice cost of $200 per line (p. 35). Another estimate is that annual operations and maintenance costs are about approximately 30 percent of capital costs (which produces cost numbers similar to the ITU estimate, if capital costs are about $1,000 per line). Thus revenues would need to exceed this amount to be profitable, and the payback period would be about three years (Kayani & Dymond, 1997, p. 51).

Reducing Infrastructure Costs in Rural Areas

It is typically assumed by both operators and regulators that the costs of providing telecommunications in rural areas are unavoidably high, and, coupled with low demand, render rural services necessarily unprofitable. While costs per line are bound to be higher than in urban areas, creative strategies for design and implementation may reduce costs.

Topography and climate are important considerations in system design. A microwave network may be an appropriate solution for plains and valleys, but satellite service is likely to be more suitable for mountainous areas. Designing for available transportation facilities and labor can also reduce costs. For example, the Alaskan carrier GCI specified that VSATs built for operation in Alaska villages must be designed to be flown into villages in small aircraft, as there are no roads to most villages. Maintenance and troubleshooting are done by bush pilots who regularly fly into the villages. Bell Canada trains local technicians to do basic telephone installation and troubleshooting in northern Canadian communities. Other strategies such as the use of prepaid stored-value telephone cards can save time and money by eliminating the need to collect coins from pay phones (while also preventing pilferage).

Modular design that allows for adding capacity when required will also reduce costs of upgrading service. Demand may increase not only with population growth, but also if there are changes in the economy or demands for new service, such as Internet access. A digital microwave system installed in the Australian Outback reached capacity much earlier than expected not only because of its design (which required remote switching for village-to-village traffic) but also because of unanticipated demand for fax and then Internet access. Upgrading the network required a complete overbuild. In the Marquesas in the South Pacific, satellite earth stations have been in-stalled for telephone service and TV reception, but circuit capacity is very limited. When asked whether additional capacity could be added if demand increased (for example, for Internet access for schools), a site engineer said, "There will never be more demand here."[12] Never assume never.

Providers of facilities and services can also use several strategies to mini-mize operations and maintenance costs, which are generally reflected in rates charged to customers. One approach is to specify equipment de-signed to function in developing country environments. For example, equipment must be built to withstand climatic conditions such as extreme variations in temperature, high humidity, dust, and sand. Satellite termi-nals and switches can also be designed to be remotely monitored, so that centrally located technicians are aware immediately if an earth station goes offline.

A second strategy is to rely on local labor as much as possible. Rather than sending out technicians with high travel and labor costs, operators can train local people to do installations, routine maintenance, and trouble-shooting. These two strategies can be combined, so that a centrally located technician may be able to advise a local person on troubleshooting. Also, systems designed to allow unskilled people to pull out and replace circuit boards or other faulty modules can reduce the need for onsite expertise. Another strategy is to install facilities such as public phones and computers in kiosks, shops, cafés, etc. where the owner has an incentive to take care of the equipment and protect it from vandalism.

Reducing End User Equipment Costs

The cost of computer equipment and peripherals can also be daunting in developing countries. Elimination of government import duties on infor-mation technology can reduce the cost of equipment for developing coun-try users and suppliers. Relying on income from duties is a short-sighted strategy anyway; the evidence cited in early chapters suggests that utilization of ICTs is likely to contribute much more to the economy, and thus to the national treasury, than any import duties. In many cases, costs can be re-duced further if local industries assemble equipment for their own markets using imported components, a strategy common in many Eastern Euro-pean, Asian, and Latin American countries.

Countries such as China, India, and Brazil are also encouraging research and development of affordable and reliable devices that provide services needed in developing regions. For example, the "simputer," designed in In-dia, is intended to be an inexpensive network access device that does not re-quire literacy, but relies primarily on images and audio. The target cost for the simputer is about $200; shared access is to be available through prepaid smart cards.[13] Brazilian researchers at the Federal University of Minas Gerais have designed a "Volkscomputer," a low cost but fully functional computer designed to sell for about $300. The government plans to install the stripped-down machines in public schools and sell them to low-wage earners on installment for as little as $15 a month.[14]

Such an approach to providing credit schemes for ICT purchases could be the most effective strategy of all. In Uganda, MTN (a mobile operator) is working with the Grameen Foundation to introduce microcredit channels that would enable more Ugandans to purchase cell phones by taking out small loans and repaying them over several months. An Egyptian official points out that many Egyptian families could afford $20 per month for a computer if credit were available. In the United States, PeoplePC has pioneered lease/purchase packages for low-income families.[15] Innovative lease/purchase schemes could make computers much more affordable, and perhaps significantly reduce the gap in computer availability between poor and industrialized countries.

TECHNICAL SUPPORT FOR ICT PROJECTS

Various approaches have been adopted to provide support for technical troubleshooting and equipment maintenance and servicing for installations such as school computer networks and community telecenters. There is no one ideal model; local expertise, budgets, and logistics will determine which among the following options (or possibly others), all of which have been used in field settings, may be most suitable for a particular project. However, what can be generalized is that technical support will be required!

• *Hiring local technical staff:* A few telecenters, such as some of the early European telecottages, chose local managers with technical expertise. However, such skilled people may not be available or may not have the organizational and outreach skills required for community access or entrepreneurial expansion of telecenters.

• *Shared dedicated technical support staff:* Some large projects that involve several locations may hire their own dedicated technical support staff whose jobs include on-demand troubleshooting as well as regular visits to each site to provide maintenance and training.

• *Technical support organization:* A related organization may have technical expertise to support several community access projects. For example, in northern Canada, K-Net provides technical support and training for several remote community telecenters (see chapter 5). Similarly, Wawatay, another native communications organization in Northwestern Ontario, provides technical support for community radio stations and two-way radios used in the bush on traplines and at fishing camps.

• *Contracting local experts:* In some cases, a local person in a technical business or a teacher, government worker, contractor, etc. may be contracted to do maintenance and troubleshooting, and possibly additional technical training.

• *Commercial contractor:* A company with the required expertise such as a telephone company, another utility company, or electronics contractor may be contracted to provide technical support. In Alaska, some local telephone companies provide technical services for telecenters, while a long-distance company has developed a specialized "one stop shop" to provide technical support in addition to connectivity for schools, libraries, and clinics.

• *Regular visitor:* A person with technical expertise who frequently visits the community may be contracted to provide technical support if there is no one available locally. As noted above, an Alaskan VSAT operator relies on bush pilots who regularly fly into the villages to provide some VSAT maintenance and troubleshooting.

• *Online support:* In addition to the above resources, project staff themselves are likely to become important sources of support to each other. There are many examples of online troubleshooting through telecenter discussion groups, as well as individual advice and support by telephone or e-mail.

IS THE FIXED NETWORK FACING EXTINCTION?

Fixed networks face numerous threats ranging from bad management and monopolistic complacency to threats to their revenues from wireless carriers and VOIP. Given these problems and the cost advantage of wireless in regions without existing infrastructure, developing countries have an opportunity to leapfrog wireline solutions such as DSL and cable to wireless broadband.

One option would be to upgrade existing networks to 2.5G (or 2.5 generation) facilities, typically GPRS for the GSM networks that are prevalent in most of the developing world. This approach builds on existing infrastructure, and is much less costly than full upgrades to 3G networks. As described above, recent technological innovations offer several other wireless possibilities. WiFi (IEEE 802.11 standards) is a low-cost means of covering small villages or public spaces. WiMAX (IEEE 802.16) may be attractive because it covers a larger area and does not require line of sight for transmission. Another option is CDMA 450 that can be deployed in cells of 50 to 60 km in diameter, and can provide data speeds up to 2.4 Mbps, making it suitable for rural areas as a fixed wireless deployment for voice and alternative to DSL or cable for Internet access. In addition, satellite links may be the most practical means to reach truly isolated communities and to provide the backhaul links for various wireless last mile (or first mile) technologies.

Is wireless ringing the death knell of traditional fixed wireline networks in the developing world? It seems likely that not only voice but also higher bandwidth services will be available primarily through wireless networks in poorer countries. The fixed wireline network may first face extinction in the developing world.

Technology Lessons for ICT Projects

The foregoing discussion refers primarily to network planning, installation, and ongoing maintenance. At the project level, several strategies may also reduce equipment costs and prolong the life of ICT facilities. The following lessons are based on field experience from numerous projects:

- Select equipment that can best tolerate local conditions (e.g., temperature extremes, lack of air conditioning, humidity, dust).
- Protect equipment as much as possible. Require use of power surge protectors to combat irregular current and spikes from lightning. Some African telecenter staff also made cloth covers to protect computers and printers from dust when not being used.
- Plan for provision of spare and replacement parts. In terms of replacement, toner cartridges for printers and copiers are likely to be most critical. Without them, users cannot make copies or printouts, an important function in schools and significant (and sometimes primary) source of revenue for many telecenters. Some telecenters in Africa did not have enough cash on hand to buy replacement toner, even though they knew that the revenue from printing and copying would more than cover the cost.[16]
- Select equipment based on anticipated local demand. For example, in several African telecenter projects, a combined printer/scanner/copier was provided, based on assumptions of needs for all of these functions. However, it turned out that demand for copying was higher than for scanning and faxing, and the slowness of the combined machines and expense of toner made dedicated copiers more practical and cost-effective.
- Involve project participants in equipment selection. In Bulgaria, each telecenter management team was given some discretion in determining the mix of subsidies and equipment provided within the budget. Based on their business plans and facilities already available, some chose more support for Internet connectivity, while others chose more computers or a different mix of equipment (Tifft, 2002).

- Consider replacement options. What happens if the equipment in the school is stolen or vandalized? If the telecenter burns down? Insurance may be advisable, but some insurers may refuse to write policies or may charge very high premiums or deductibles for what they consider to be high-risk locations. If insurance is not a viable option, the project should anticipate responsibility for damage and losses: Are the costs to be borne by the telecenter? By the community? By the agency initiating the project? Or other sources or some combination of these?

- Prevent viruses! Computer viruses from online sources and local sources such as diskettes and other computers can wreak havoc with school networks and telecenters. Computers must be equipped with virus scanning software and subscriptions for online updates. Staff must also be trained in the importance of preventing viruses from getting into their computers and in dealing with infected diskettes and hard drives. Viruses caused many problems and significant downtime at African school networks and telecenters.

NOTES

1. Most of the developing world uses the 2G or second generation GSM standard, developed in Europe. The other major standard, developed in the United States, is CDMA (code division multiple access).
2. See www.gsmworld.com
3. See www.wiMAXforum.org
4. See www.cdg.org/technology/3g/cdma450.asp
5. For more information on wireless technologies for developing regions, see M. Jensen (1999), "The Wireless Toolbox: A Guide to Using Low-Cost Radio Communication Systems for Telecommunication in Developing Countries—An African Perspective." http://web.idrc.ca/es/ev-10592-201-1-DO_TOPIC.html
6. See www.direcway.com
7. See www.gilat.com, www.tachyon.net, and www.vitacom.com
8. See www.direcway.com
9. See www.firstvoiceint.org and www.worldspace.com
10. See www.vita.org
11. See, for example, www.freeplay.net
12. Personal interview by the author, December, 1996.
13. See www.simputer.org
14. See www.bridges.org/spanning/chpt2.html
15. See www.peoplepc.com
16. Personal observations from field work by the author for the IDRC Acacia Initiative.

Chapter 8

Restructuring the Telecommunications Sector in Developing Regions

A competitor in a developing country remarked that when he called the Ministry to get a draft of the new Telecommunications Act, the secretary said she couldn't give him a copy because she hadn't received the diskette from [the incumbent] yet.
—Personal interview by the author (2004)

This chapter addresses strategies to restructure the telecommunications sector in developing regions. Although some researchers have argued that competition is not viable in poorer countries or regions, or that competition would not increase access to telecommunications among lower income populations, this chapter argues that restructuring to allow competition, within a framework of policy goals and with effective regulatory oversight, can significantly improve access. As a case in point, this chapter also examines lessons from the phenomenal growth of wireless in the developing world, where, as noted earlier, in many developing countries there are now more wireless than fixed lines, and for many subscribers, their cellphone is their first and only phone. In most of these countries, competition has been allowed in wireless, and the result has been lowering of prices and extension of coverage. Other policies and strategies designed to extend access within a competitive framework are also discussed.

THE NEED FOR LIBERALIZATION

The newly industrializing economies of eastern Europe, Asia, and Latin America are starting to close the gaps with industrialized countries in terms of access to basic telephony and the Internet. Their growing economies appear promising to the telecommunications industry and to investors who are looking for new markets. Most of these countries are also taking steps to encourage investment by privatizing their operators, providing investment incentives, and/or introducing competition.

Information gaps show least signs of shrinking in the poorest countries. Yet, as these countries develop market economies and seek to take maximum advantage of scarce expertise, they will need to invest in telecommunications. Of course, these regions are less attractive to investors than more prosperous economies; in general, they have also been the most reluctant to reduce their governments' role as monopoly operator. Their networks are also the least efficient, in terms of reliability and the number of lines per telecommunications employee. Restructuring their telecommunications sectors to improve productivity and encourage investment will be necessary if they are to begin to close the gap.

Competition is likely to offer the greatest stimulus for innovation in technologies, services, and prices. Competition can range from terminal equipment (now competitively available almost everywhere) to new services such as cellular telephony, to value-added services such as Internet access, to full competition in all network services. Most developing countries have opened provision of Internet services to competition; many have licensed more than one mobile wireless operator. Competitive ISPs typically offer lower prices and more variety in service packages than a monopoly ISP run by the telephone carrier. However, the ISPs are dependent on telecommunication carriers to reach an Internet gateway. Where fixed services are still a monopoly and bypass is not allowed, the ISPs are faced with high costs which are passed along to their subscribers. Local monopolies that charge users by time and/or distance add to the costs of using the Internet.

Privatization of government-owned PTTs is usually the first step in the restructuring process. But privatization alone is not sufficient. Private monopolies may be run in a more business-like manner than government entities, but they may be content to rely on "cream-skimming" to serve businesses and the most affluent individuals rather than lowering prices and extending services to other potential customers. Only when competition is introduced will there be incentives to adopt strategies to cut costs, and to target new markets. Where competition has been introduced, the results can be dramatic.

THE IMPACT OF COMPETITION

What has driven the explosive growth of wireless in developing regions, despite the anemic growth of fixed networks? It appears that the key driver is competition. Wireless competition has resulted in innovative pricing and service offerings. Rechargeable smart cards make phone service accessible to people without bank accounts or credit histories. Cheap text messaging can substitute for many e-mail functions.

Low- and lower-middle-income countries with at least two wireless providers have significantly higher wireless growth rates than countries with a wireless monopoly. (see Fig. 8.1). Figure 8.1 compares wireless growth rates in 79 developing countries that have at least two mobile operators with growth rates in 15 developing countries that have a mobile monopoly. Countries with wireless competition are not wealthier than low-growth-rate countries; in fact, both categories include some of the world's poorest countries, such as Cambodia, Indonesia, and Bangladesh in Asia, and Benin, Burkina Faso, Rwanda, and Uganda in Africa.

INVESTMENT INDICATORS

Investment in the telecommunications sector in poorer countries has lagged investment in wealthier countries. While investment per subscriber is comparable, the number of subscribers in most developing countries is relatively low. A more useful comparison is investment per inhabitant (see Fig. 8.2). Here we see that investment in the poorest countries is only $3.3 per inhabitant, and $18 per inhabitant in lower middle-income countries, compared with $126 per inhabitant in high-income countries. Note that these figures apply primarily to fixed networks.

THE REVENUE GAP

Of course, we would expect revenue to be lower in absolute terms in lower income countries, but in fact revenue averages about 3.1 percent of total GDP in all but the poorest countries, where revenues are just 1.8 percent of GDP (see Fig. 8.3).

What would be the financial impact if telecommunications revenues in the poorest countries could be increased from 1.8 percent to 3.1 percent of their GDP, the average for all other nations? (Lower-middle-income countries actually generate revenues of 3.4 percent of GDP.) The total population of the low-income countries is about 1.1 billion. Thus, such an increase in revenue would generate an extra *$14.3 billion* per year!

Drawing again on the lessons of wireless growth, it appears that better marketing including pricing and service options could significantly contribute to generating funds to close the digital divide. The following strategies and policies could also create incentives to generate more revenue, reducing the need for subsidies.

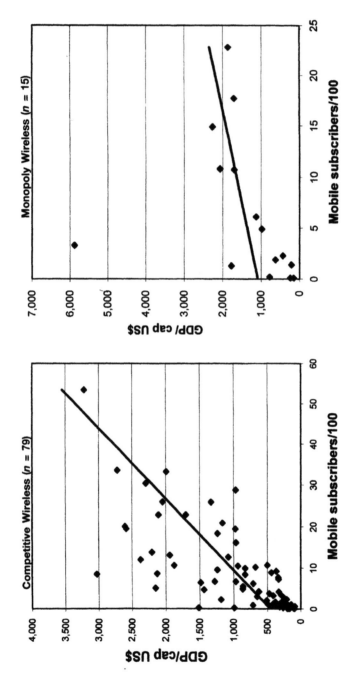

FIG. 8.1. Wireless cellular growth rates in low- and lower-middle-income countries: monopoly vs. competition. Derived from ITU data for 2003.

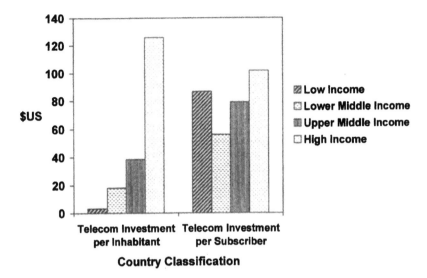

FIG. 8.2. Telecommunications investment. Derived from ITU data for 2003.

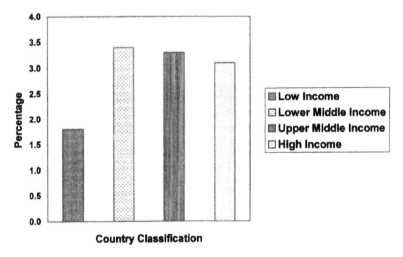

FIG. 8.3. Telecom revenue as percent of GDP. Derived from ITU data for 2003.

IMPROVING TELECOMMUNICATIONS MANAGEMENT

Why might investment in fixed networks be so low, given the evidence of pent-up demand shown by the growth of wireless networks? Most fixed networks in developing countries remain monopolies, even in countries that have made some regulatory commitment to introducing competition. Well

TABLE 8.1
Management Indicators

Country Classification	Faults/100 Lines per Year	Subscribers per Employee	Revenue per Employee
Low Income	105.1	113	$21,624
Lower Middle Income	20.8	243	$73,568
Upper Middle Income	15.1	382	$157,243
High Income	10.5	453	$299,356

Source. Derived from ITU data for 2003.

managed monopoly networks could provide good opportunities for investment. However, these networks are generally very poorly managed. Consider reliability. Fixed networks in low-income countries had 10 times as many faults per 100 lines as networks in high-income countries. Thus many of the limited number of available lines were likely not to be working properly and not generating revenue.

Also, management is much more efficient in wealthier countries, which have about four times as many subscribers per employee as do networks in the poorest countries. Some of this difference may be due to the lower price of labor in low-income countries, but another factor is likely the tradition or requirement of using the telecom operator as a source of jobs in countries where there is still significant government ownership in the telecommunications sector.

Another important management indicator is revenue generated per employee. Not only is absolute revenue greater in high-income countries, but revenue per employee is more than 13 times as much as revenue per employee in the poorest countries. See Table 8.1.

LIMITING PERIODS OF EXCLUSIVITY

In a liberalized environment, the length and terms of operator licenses can impact the pace of growth of networks and services. Regulators typically face choices concerning how long to protect incumbents to enable them to prepare for competition, and how long to grant periods of exclusivity or other concessions to new operators to minimize investment risk. Yet exclusivity and long time periods may be the wrong variables to focus on if the goal is to increase availability and affordability of telecommunications services. Instead, investors cite a transparent regulatory environment with a "level playing field" for all competitors and enforcement of the rules as key to their assessment of risk. It is highly unlikely that fixed-line providers will have an incentive to roll out broadband services beyond large businesses and some upscale residential areas if they see no near-term threat to their monopoly. Some jurisdictions[1] have negotiated terminations of exclusivity periods with monopoly operators in order to enable their economies to benefit from competition in the telecommunications sector.

EXTENDING ACCESS THROUGH RESALE

Authorization of resale of local as well as long distance and other services can create incentives to meet pent-up demand even if network competition has not yet been introduced. Franchised payphones can be introduced in developing countries in order to involve entrepreneurs where the operator has not yet been privatized and/or liberalized. Indonesia's franchised call offices known as Wartels (Warung Telekomunikasi), operated by small entrepreneurs, generate more than $9,000 per line, about 10 times more than Telkom's average revenue per line. In Bangladesh, Grameen Phone has rented cellphones to rural women who provide portable payphone service on foot or bicycle to their communities. Franchised telephone booths operate in several African countries; in Senegal, private phone shops average four times the revenue of those operated by the national carrier (ITU, 1998, pp. 77–78). In the Philippines, entrepreneurial wireless subscribers resell small increments of time, transferring them to "pay as you go" customers who cannot afford the price of a phone card.

Resale of network services can also reduce prices to customers. Most interexchange carriers in industrialized countries are actually resellers that lease capacity in bulk from facilities-based providers and repackage for individual and business customers, offering discounts based on calling volume, communities of interest, time of day, and other calling variables. Similar strategies can be used to resell broadband when networks that are upgradeable (such as for DSL) or that have excess capacity (such as optical fiber or satellites) are available.

LEGALIZING BYPASS

Facilities that are not offered by an incumbent wireline operator may be the least expensive means to extend access. For example, a VSAT may be an ideal solution to bring high-speed Internet access to a rural school or telecenter. A WiFi "hotspot" may be a low-cost means of providing broadband to a village or neighborhood. However, in some developing countries, even if the wireline provider does not provide broadband services in the area, or possibly does not even serve the area, such alternative connections would be considered illegal bypass.

Many monopoly operators claim that bypassing their networks effectively siphons off revenues that they need to expand their networks. However, the relationship is not so simple. As noted earlier, without competition, there is likely to be little incentive to roll out new services such as broadband, to choose the most cost-effective technologies, and to price services reasonably. Thus, policymakers will not further the goal of extending access to affordable broadband and other services by preserving wireline monopolies.

Protecting dominant carriers may also hinder economic growth. For example, a West African Internet service provider pointed out that he needed relatively inexpensive international connection to the Internet in order to provide affordable Internet access for his customers. By using bypass, he is creating new jobs in value-added services as an Internet provider, as well as providing an important information resource for economic development of the country.[2]

REDUCING LOCAL BARRIERS

Some developing countries set duties on imported equipment including computers and components for telecommunications networks. These duties may generate revenue for the government, but they increase the cost of network facilities and end user equipment, and thus increase the cost of access for its citizens. The economic benefits of having available and affordable access are likely to outweigh substantially the value of such fees and duties.

Governments may also inhibit network buildout by making it difficult for operators to secure permits for rights of way or use of existing poles or conduits, or by charging fees for such permits or other services that place a significant financial burden on the operator. While such fees may also be attractive sources of revenue for the government, they may have the effect of delaying access to the Internet for its residents. In the United States, the FCC and some states such as Michigan are working to reduce local barriers in order to facilitate buildout of broadband networks.[3] Michigan's plan calls for a 45-day turnaround to process rights-of-way permits and eliminates redundant charges if a provider wants to offer more than one service on its lines, such as cable Internet access as well as cable video.[4]

LOCAL FRANCHISES

Although in most countries a single carrier provides both local and long-distance services, it may be advantageous to delineate territories that can be served by local entities. In the United States, the model of rural cooperatives fostered through the Rural Utilities Service (formerly Rural Electrification Administration) has been used to bring telephone service to areas ignored by the large carriers. As noted above, wireless technologies could change the economics of providing rural services, making rural franchises much more attractive to investors. Local enterprises are likely to be more responsive to local needs, whether they be urban or rural. An example of this approach in urban areas is India's Metropolitan Telephone Corporation, established to serve Bombay and Delhi. Local companies also provide telephone service in Colombia. Cooperatives have been introduced in Hungary and Poland. A disadvantage of this approach is the need for local expertise to operate the system, which is likely to be in particularly short supply in many developing countries.

COMPETITIVE RURAL OPERATORS

Some countries have granted monopoly franchises to rural operators to encourage their investment. For example, Bangladesh has licensed two monopoly operators in different rural areas; they are allowed to prioritize the most financially attractive customers and charge a substantial up-front subscriber connection fee. The Bangladesh Rural Telecommunications Authority (BRTA) is profitable, even though it has to provide at least one public call office (PCO) in each village that requests one (Kayani & Dymond, 1997, p. 18).

However, other countries are opening up rural areas to competition as part of national liberalization policies. Argentina allows rural operators to compete with the two privatized monopolies, Telecom and Telefónica. Some 135 rural cooperatives have been formed to provide telecommunications services in communities with fewer than 300 people (Kayani & Dymond, 1997, p. 18). Finland's association of telephone companies has created several jointly owned entities that provide a range of rural, local, and long-distance services in their concession areas, in competition with the national operator (p. 19). In Alaska, a second carrier, GCI, competes with AT&T Alascom to provide long-distance services in rural and remote areas. Canada opened remote regions to competition in the year 2000.

INNOVATIVE PRICING

If demand may be greater and costs lower than is commonly assumed, why are prices of connectivity often so high in poor developing regions? Typically, the pricing reflects the structure of the telecommunications sector. As with other goods and services, where competition exists, prices tend to be lower. However, operators may increase revenues by offering attractive pricing to increase traffic volumes, particularly on underutilized networks. The following are some pricing options now being offered in some industrialized countries that could be adopted in developing countries:

- **Discounts for Off Peak Use:** Discounts are typically introduced to generate additional revenue from capacity that would otherwise be underutilized (e.g., at night and on weekends). Such rates also provide more affordable personal communications, for example, for family members away at work or at school, to stay in touch. They may also be help to make Internet access and e-mail more affordable.

- **Discounts for Government/Public Services:** There is a high social value in ensuring that public social services have access to reliable telecommunications. A discount plan for public services may be introduced where governments have limited funds available. As noted in chapter 6, the U.S. Telecommunications Act of 1996 mandates discounts for "advanced services" such as Internet access for schools, libraries, and rural health care facilities. Such discounts may also increase revenues if these institutions would not otherwise be able to afford Internet access.

- **Pricing Based on Communities of Interest:** Affordable rate structures are particularly valuable for communities of interest (i.e., locations that are called frequently such as clusters of villages where there are close family ties, provincial capitals, or major trading centers). One approach to community of interest pricing is Extended Area Service (EAS), which offers callers an option of discounted or flat rate calling within a zone. Another approach is to offer toll-free calling to regional government offices or other important social services.

- **Flat Rate Pricing:** The cost of communicating by geostationary satellite is independent of distance, as signals must travel 72,000 km to the satellite and back, whether the users are 300 km or 3,000 km apart. Even in terrestrial wireline networks, the cost of transmission has dropped dramatically as available bandwidth has increased. As a result, flat rate pricing is being introduced in many industrialized countries to replace distance-based pricing. The effect is to eliminate the distance penalty, which can be a major benefit for rural and remote users. Mali has introduced flat rate pricing for Internet access, so that it costs no more to connect to the Internet from Timbuktu than from Bamako, the capital.

- **Discounts for Unused Capacity:** Operators may offer reduced rates for capacity that would otherwise be idle. For example, some telephone companies in the United States offer "dark fiber" rates for access to optical fiber capacity that is currently not used ("dark"). While we might assume there are shortages of capacity in developing regions, there may be at least temporary gluts in network capacity where fiber backbones are installed, as is now the case in some submarine intercontinental and regional fiber networks. Discounts on this unused capacity could both generate more revenue for suppliers and provide more affordable access for users.

REGULATION AND POLICY ISSUES

Oversight With Enforcement

All of these suggested policies require oversight to monitor progress toward meeting targets, to enforce compliance with performance standards, and to review and revise benchmarks. For example, there will be a need for monitoring to determine whether there are disparities in access, quality of services, or pricing that need to be addressed. Operators must also be held to their license conditions if these terms are to be an effective means of extending access. Otherwise, operators may not meet license targets in areas they think will not be profitable, or may install facilities but not maintain them adequately if they assume the revenue-generating potential is low.

The common wisdom is that regulators must be independent both of the industry and of the political process. In countries that began with the PTT model, the concern is to make the regulator independent from the PTT to avoid conflict of interest between operator and regulator, a necessity in a competitive environment where equitable rules must be set and enforced for all operators. However, a problem with this approach is that typically employees who once worked for the PTT now work for the regulator, making it inherently difficult to avoid bias in assumptions or decisions.

It may be possible to guard against explicit bias favoring the former PTT, but perhaps a more significant danger is that regulatory employees will start with assumptions of what is feasible and practicable based on their PTT experience. For example, they may not believe that installations can be done faster or more cheaply than under the incumbent PTT, or that newer technologies may be more suitable and cost-effective. Such responses can hinder the abilities of new entrants to obtain approvals or licenses to roll out facilities that could provide new or lower cost services. Strategies that could address this problem would be to include professional staff such as economists from other ministries, use external consultants without ties to the former PTT, and request public filings and comments so that all relevant views and experience may be considered.

If the Government Is Slow to Act, Regulation Becomes Policy

A distinction is often made between policymaking, typically carried out through a government ministry or department with responsibilities for telecommunications, and regulation, to be carried out by an "independent" body (i.e., that is not related to the operator or directly responsible to a minister). However, the distinction between regulation and policy quickly becomes blurred, because of the pace of technological change and market pressures in the communications industry. Some countries[5] have made a functional distinction in placing much more limited responsibility in the regulator as an adjudicator and arbitrator, while retaining responsibility for licensing as well as policymaking within the ministry. Although perhaps attractive conceptually, a danger of this approach is that the government will not respond in a timely manner, resulting in delays for customers and lost opportunities for the communications industry. One strategy to avoid this problem is to set firm enforceable deadlines for decisions on license applications and other time-sensitive matters. This approach was used in the U.S. 1996 Telecommunications Act which set specific deadlines for the FCC to complete various rulemakings and directives required to implement the Act.

Separating the Goals From the Means

The role of government should be to set goals and not to determine how they are to be achieved. However, regulators and policymakers have a tendency to confuse the goals with the means. For example, in the United States in the 1970s, the FCC initially tried to dictate the size of antennas and technology to be used in two-way VSATs in order to minimize interference. Innovative engineers were able to convince the FCC to set the technical specifications and let the industry determine how to meet them. The result was smaller and cheaper terminals than would have otherwise been developed.

A recent example from India would be a government requirement to upgrade village telephones (PCOs) to provide data communications. Perhaps the goals should be stated as providing access to e-mail and the Internet in every village. The means to achieve the goal may range from upgrading village telephones, to establishing public access in schools or community buildings, or in privately run business centers or tea shops. The government's role would be to ensure that reliable and affordable access is available in every community.

Effective Regulation Requires Participation

It is often thought that the issues in telecommunications policy and regulation are so technical and arcane that most people would have nothing useful to contribute to the decision-making process, and that public participation would therefore add little of value. However, all regulatory agencies are overworked and understaffed, and cannot find or analyze all the data that would useful to guide decision-making. Major business users are likely to have well-thought-out views on the impact of proposed regulations or the need for reforms that would enable the telecommunications sector to better serve their industries. The ability of small businesses, NGOs, and consumers to contribute may seem less obvious; it may take some time for their representatives to get up to speed on telecommunications technology and economics. However, the contribution of such groups can also provide perspectives that might otherwise be overlooked. Organizations such as the Alliance for Progressive Communications (APC) are working to build civil society knowledge about ICT policies in Africa.[6]

A problem for consumer groups is the cost of tracking the issues and preparing testimony or other interventions. In order to ensure that such consumer perspectives are represented, in some countries[7] the regulator pays the costs of participation in hearings by consumer organizations that contribute evidence which would not otherwise be available. This approach may at first glance appear beyond the means of many regulating bodies. However, it offers a relatively inexpensive means to receive relevant information that would otherwise be very time-consuming or costly to obtain (e.g., through staff studies, hired consultants, or field research). In addition, it provides an incentive for consumer organizations to become knowledgeable and effective advocates on communications issues. For example, native organizations in Canada have frequently testified in regulatory hearings on the need for telecommunications services, and quality and pricing of existing services, in isolated northern communities.

NOTES

1. For example, Hong Kong, Singapore, and India.
2. Personal communication, July, 1997.
3. See www.fcc.gov/broadband
4. Clara Garretson (2002), "Broadband group wants unified right of way plan." PCWorld, May 30. http://www.pcworld.com/news/article/0,aid,101550,00.asp
5. For example, India has placed licensing responsibility with the Division of Telecommunications Services in the Department of Telecommunications rather than with the Telecommunications Regulatory Authority of India (TRAI).
6. See Association for Progressive Communications, www.apc.org
7. For example, the Canadian Radio-Television and Telecommunications Commission (CRTC) in Canada, and the California Public Utilities Commission in the United States.

Evaluation: Issues and Strategies

If I learn to use a computer, will I be able to get a job?
—Woman at a telecenter in South Africa

The World Bank's InfoDev initiative points out the need for rigorous research on ICT projects:

> [R]igorous field-tested knowledge about "what works and why" in ICT for development, and a deeper understanding of the enabling conditions and success factors in ICT-for-development initiatives, have been relatively scarce. As a result, there is a growing consensus in the development community that ICT will only become an effective and mainstream tool of poverty reduction and sustainable development if the proponents of ICT-for-development can provide more rigorous evidence, strategies, benchmarks, indicators, and good practices that are directly relevant to the core poverty-reduction and development priorities of developing countries and their international partners.[1]

Indeed, few of the ICT projects described in earlier chapters have had rigorous evaluations. This chapter addresses the challenge by proposing evaluation frameworks and methods for ICT field projects. Telecenters are used as the primary case because of their proliferation in developing regions, largely through support of international organizations and foundations. Evaluation techniques are then suggested for ICTs in education and for telemedicine projects in rural and developing regions.

Telecenters are intended to meet the evolving information, communication, and learning needs of their communities. It is therefore important to make a systematic effort to monitor the costs and usage patterns and to measure the impact of telecenter usage. As Scharffenberger (1999) observes, the need for monitoring and evaluating telecenter efforts has been raised in multiple conferences and meetings, but unfortunately the rhetoric has not always been matched with the necessary energies and resources. Budget allocations for monitoring and evaluation are often modest or nonexistent and all too often, those managing the projects end up reporting that setting up the centers consumed all of the available time and money.

THE PURPOSES OF EVALUATION

Evaluation may serve several purposes. *Formative* evaluation relates to process. It can provide feedback on individual sites—for example, to identify strengths and weaknesses in an inaugural telecenter to provide guidance for a development agency for its next installations—and on multiple sites, how well these are working, what changes or improvements should be made, and what was learned that could be applied in other similar projects. This information can also be useful to other agencies supporting similar projects.

Summative evaluation relates to outcome and impact: What difference did it make? For example, did the project achieve its goals? What was learned about the contributions of ICTs to social and economic development? These are usually the most important questions for funding agencies concerned about whether their investments have made a difference, and for decision makers responsible for budgeting for ICT facilities and services or establishing universal access policies.

One way to approach summative evaluation is to determine to what extent the project achieved its goals. This approach assumes that the project started with explicit goals, or goals that could be easily made explicit from the project plan, and that the goals can be explicated as concrete targets that can be measured or tracked. Identifying goals is further discussed below. Another way of approaching summative evaluation is to consider *who* benefitted, directly or indirectly, from this access to information or the means to create and share it. If specific target groups have been identified, such as women, young people, local NGOs, entrepreneurs, artisans, etc., answers can be sought to such questions as how have each of these groups benefited, and what might be the cumulative effects of such benefits on the wider community?

The purposes of formative and summative evaluation are related in that the feedback or process information may help to improve projects so that they are more likely to accomplish their goals, and some of the data collected about who is using the facilities and for what purposes, can be useful both for feedback to the project and for tracing the impacts of projects, as shown next.

INFORMATION AND THE DEVELOPMENT PROCESS

In designing the evaluation of ICT projects, it is necessary to consider the role of information in the development process. ICTs are not simply a connection between people, but a link in the chain of the development process itself. As discussed in chapter 2, the ability to access and share information can contribute to the development process by improving effectiveness, equity, efficiency, and reach.

However, none of these benefits occur in isolation. Those who need to access or share information must have the necessary skills or access to a resource person to help them. Other factors may also need to be considered if people are to benefit from the information, for example, access to credit for inventory or spare parts, a transportation system to get goods to market, or a curriculum that can be adapted to new teaching techniques and information sources. Thus, in order to learn how telecenters may contribute to development, we must not only find out whether and how they provide access to tools to create, access, and share information, but also identify what information is needed in their communities, and what other factors may influence the initiation and impact of activities that may be considered developmental from economic, social, cultural, and/or political perspectives.

THE CHAIN OF INFERENCE

Many of the developmental goals proposed for ICT projects imply causality between their use and specified outcomes. This chain of inference must be made explicit to trace any causal connection between provision of the facilities and development. Such a chain of inference for telecenters may be complex, because they are typically intended to serve a variety of community needs, which may not be as clearly defined as in projects designed for particular sectors or target groups. For telecenters to have an impact on development, the following at least are required:

- Community access:
  - The equipment must be conveniently located.
  - The telecenter must be open at hours when people want to use it.
  - The services must be affordable to the target groups.

- Awareness:
  - The community members must be aware of the telecenter and the services it offers.
- Skills:
  - The community members must be able to use the equipment or to get assistance in doing so.
- Lack of barriers:
  - There must be no constraints that would unduly hinder the impact of utilizing the ICTs (e.g., lack of jobs or entrepreneurial activities, cultural norms that affect certain groups such as women, lack of transport to reach new markets, etc.).

Some benefits may accrue to individuals using the telecenter such as getting help in emergencies by contacting a doctor or saving time by using telecommunications to arrange transport logistics or to substitute for travelling to the city. Other benefits may require more complex types of information-seeking. In thinking about users and potential benefits, two concepts from diffusion theory of innovations and the impact of communications may be relevant:

- *The two-step (or multi-step) flow model:* The user may not be the real beneficiary. For example, the user may be the agricultural extension agent, the health worker, the social worker; while the beneficiaries are the farmers, women and children, disadvantaged groups, etc.
- *The early adopter:* Some people may be more likely to use ICTs sooner than others, for example, those with more education and those with clearly defined information needs such as teachers, community leaders, artisans who need to find markets, and merchants who need to contact suppliers (Rogers, 1995).

SHORT-TERM VERSUS LONG-TERM IMPACT

One of the dangers in the recent enthusiasm about the role of ICTs in development is unrealistic expectations of significant short-term impact among donors and policymakers. There may be some dramatic cases of the value of access to information, for example, farmers getting better prices for produce, artisans finding new markets, health workers receiving assistance and saving lives, etc. But much of the impact is likely to take longer and be much more indirect. Better access to more up-to-date information *about planting methods* may eventually result in better yields and thus more income for farmers, and schools with access to the Internet may pro-

duce more graduates prepared to continue with their education or to qual-
ify for jobs. But these effects take time!

While planning evaluation that can capture these longer term benefits,
researchers should look for evidence of ICT usage that could lead to lon-
ger-term impact, such as women who have learned new skills, NGOs that
have been able to obtain relevant information, and entrepreneurs who
have obtained information about new markets, etc. These are more likely to
be the kinds of changes that can be documented in the first year or two of a
project's operation. Evaluators should be able to document these changes,
and identify the impacts that could result over time such as new jobs or
more trade, and the barriers that may impede longer term impacts, for example, lack of funds to continue the project, lack of credit to buy recommended fertilizers or pesticides, difficulties in employing newly trained
workers, and so forth.

EVALUATION AS LEARNING

It is most important that evaluation not be perceived as judgmental by the
participants but as an opportunity:

- To provide feedback to project staff on what is working well and what
  needs to be changed or improved;
- To plan for the sustainability of the project;
- To identify successful strategies and lessons learned that could be
  shared with other projects or networks.

Evaluation can appear threatening to field staff who may resent someone
looking over their shoulders or fear that they are being judged. One way of
reducing such fears is to suggest to the staff that what they have learned
would be useful to others starting a similar project, by asking them, for example, "What problems did you face and how have you solved them, or
tried to solve them?" and "What advice would you have for someone else
starting a project like yours?"

Another useful strategy is to involve the staff in identifying what information would be useful to them and then, rather than relying on outside researchers, training these staff and the user groups to be interviewers and
data collectors. In this way, the participants themselves hear from the community and collect information on how many people are using the telecenter (for example) and for what purposes and what their views are on the
services. The danger in this approach is that the data collection may be bi-
ased if the participants "tune out" any negative feedback. However, if they

approach the data collection as a learning process, and if they are trained in asking questions consistently and coding the responses accurately, the evaluation will be much more valuable to them. For example, in the Acacia baseline studies of telecenters in Mali and Uganda, telecenter staff were trained in interviewing and sat in on focus groups, hearing directly from community members about their information needs and perceptions of the telecenter. Similarly, any ongoing monitoring of activity such as usage logs for telecenters should be designed in consultation with the staff, showing them how the information collected can be useful for them in monitoring trends in usage and understanding the users' needs.

IDENTIFYING OBJECTIVES

In order to plan the evaluation, it is important to understand the objectives of the project. This may sound straightforward enough, but the various stakeholders such as donors, partners, communities, and staff may have quite different objectives. Also, the objectives may be rather vague or general, such as empowering local people or creating new economic opportunities. In addition, reasonable timeframes for achieving different objectives may vary. For example, teaching young people to use computers may be accomplished much more quickly than generating new jobs for the community.

To identify the objectives, evaluators may be able to obtain information from existing sources such as project documents. Where such sources are lacking or unclear, they may need to interview representative stakeholders to ascertain their objectives for the project. However, evaluators often find that the objectives stated in documents or interviews need to be clarified and made explicit in order to make decisions on methodologies, variables, and instruments. For example, the Acacia project document stated that Acacia was designed "to empower sub-Saharan African communities with the ability to apply information and communication technologies [for] their own social and economic development." Evaluators had to work from this general goal to identify specific objectives for particular Acacia ICT projects.

One approach to identify objectives is to ask the various stakeholders: "What would make this a successful project?" Their answers might include:

- Access to ICTs for people in the community;
- Community residents trained in use of ICTs;
- Usage of the telecenter by target groups: e.g. women, youth, entrepreneurs, etc.;

- Increased awareness of the importance of information in local development;
- The ability of the project to continue to operate past the pilot phase;
- Economic development in the community (e.g., job creation, better prices for products, new outlets for products from the community, etc).
- Social development in the community (e.g., the adoption of practices to reduce disease such as better sanitation, maternal and child care, improvements in community basic literacy or school completion rates, and new job skills).

All of these goals have been proposed for telecenter projects by various stakeholders.

Having made these goals explicit, the evaluators must then devise methods and tools to determine to what extent the project achieved them. Questions they need to consider include: What does access mean? How should we define sustainability? How can we isolate the effects of the project from other factors that might influence economic and social development?

RESEARCH DESIGN: BEYOND ANECDOTES

Many of the numerous sources on telecenters (including Web sites, conference papers, and reports from field visits) contain stories and anecdotes that provide useful insights and lessons learned from individual projects. While interesting, stories alone may lead to unsubstantiated conclusions or overgeneralizations. Consider the following statements about telecenters:

Women are more likely to use telecenters if telecenter staff are women (or include women).

The business model makes a difference in development impact (e.g., a business-oriented telecenter is likely to contribute less to social and/or political development than one with explicit development goals).

The skills and attitudes of the telecenter staff make a difference in developmental impact (e.g., a person trained in tracking down information or a person with community outreach skills may contribute to more developmental use of the telecenter).

Training a core group of users results in more usage of ICTs (or faster take-up by target groups) than a drop-in self-teaching approach.

All of these are assumptions that may be based on experience at one or more sites, but have not been broadly substantiated. These could, however, be formulated as hypotheses to be tested using a research plan designed to control for extraneous factors.

TESTING HYPOTHESES: RESEARCH DESIGN

Much of the research on ICT projects is based on case studies. Well-designed case studies can be invaluable in understanding the experiences and lessons learned in particular communities. However, they may not address fundamental questions of causality such as, for example, whether the telecenter actually contributed to the creation of new jobs in the community, or whether lessons learned could be generalized to similar telecenters in different locations or to different types of telecenters. Where there are several project sites or opportunities to track a variety of sites over time, it may be possible to gauge longer term impacts and issues of causality through research designs which are known as "quasi-experimental" because in the field settings they cannot control all the extraneous factors as can be done in laboratory settings. The following designs are not perfect, but are superior to stand-alone case studies in improving the *validity* and *generalizability* of findings:

- *Before-After:* Collecting data on specified indicators before and after the installation of the telecenter;
- *After Only:* Where no baseline information is available, it is difficult to isolate and quantify impacts. However, strategies that can be used here include retrospection, that is, asking the users to state how they got the information or carried out the task before the telecenter, and contrary-to-fact questions such as: If you did not have the telecenter, how would you do this?

The danger in both of these designs is that they can lead to false conclusions such as "the telecenter created more jobs" because they do not control for extraneous factors that might have had this effect anyway, such as another development initiative or a new road. Validity and generalizability can be improved by adding a control group.

- *Matching:* Evaluators can gain better insights into causality if they can add a group of sites that are similar in population size, isolation, economic base, etc., and collect the same data at all of these sites.
- *Random selection:* It may be possible to use randomly selected sites if there is a large data set to draw from, or if the project can be designed to randomize choice of where the telecenters will be installed. However, this approach is quite rare and can only be conducted where

there are a large number of sites such as schools with computers, phone shops, or cybercafés, or with a large existing database such as the one generated by a baseline study of rural communities in South Africa.

- *Controls using various forms of telecenters:* In countries where there are phone shops with ICTs, private telecenters, cyber cafés, etc., these sites could be included in the sample to test the hypothesis that the business model makes a difference in usage and benefits.

- *Multiple measurements:* Whether or not a control group of sites can be included, collecting data at several points after the telecenter is installed is likely to provide better insights into causality than any single "After-Only" data collection. Also, later waves of data collection will help to determine whether usage of the telecenter dropped off after initial interest, or whether demand and applications changed over time.

- *Sampling:* In collecting community data, using a systematic approach to drawing a sample (rather than interviewing the first people encountered or people known to the interviewers) strengthens the validity of the data. There are many approaches to drawing samples that are appropriate for ICT project evaluation (see Hudson, 2001).

SUSTAINABILITY

The evaluation should include an assessment of the sustainability of the activity past the pilot project phase. Of course, findings on benefits and impacts will be important here. If the project has not had much impact on the target population or has failed to achieve its primary objectives, its future sustainability may be of little consequence. However, where projects have achieved their objectives or are well on the way to doing so, it will be important to collect and analyse data that can provide an indication of future viability, for example:

- Costs and Revenues:
  - What revenues does the project generate now? Are these markets likely to remain stable, increase or decrease? (Reasons for the latter might include installation of additional public telephones in the community, additional providers of some ICT services, and the purchase of computers and modems by the wealthier individuals in the community or the business patrons.)
  - What were the start-up costs (e.g., equipment, building, training, etc.)?

- What are the on-going operating costs of the project (e.g., personnel, supplies, spare parts, rental, utilities, technical support)?
- Are the projected revenues sufficient to cover the on-going costs?
- What other sources of revenue might the project develop, for example, selling additional services, finding major clients as underwriters, and building operating costs into an organizational budget?
- Are there ways of reducing costs (e.g., through discounts from telecommunications operators, or sharing equipment or staff)?
- Facilities:
  - Has the ICT equipment proved reliable in field settings?
  - Are the power supplies and telecommunications networks sufficiently reliable?
  - Is timely technical support available when needed?
- Staff:
  - Do the current staff have the skills to continue to operate the facilities?
  - Is the current project management committed to continuing with the project?
  - Are there others who could be recruited and trained to work on the project?
- Commitment:
  - Is the activity a priority for the target group(s) (e.g., school district, health ministry, community)? The project may die or be left in obscurity, or equipment may be vandalized, if there is not a strong commitment from its constituents.

Such analysis is important, regardless of whether the project is assumed to be viable as a stand-alone enterprise by the end of the pilot period, or whether it is expected that ongoing support from donors or government agencies will be required. For example, even if the equipment was donated, it is important to monitor the operating costs in terms of repairs and spare parts and to estimate a realistic depreciation schedule to determine when it will have to be replaced. If buildings or other utilities are donated, it is also important to estimate their real costs to ensure that these are included in estimating on-going operating costs. If staff are paid by project funds, it is also important to estimate their costs in terms of salaries and benefits; the value of volunteer assistance should also be calculated.

This information is valuable not only to the donors in estimating the real costs of an operational project such as a telecenter, but also to the project managers. It helps them determine what revenue will be required to sustain the project and what strategies are needed to generate that revenue—sales of services alone or a mix of sales, donations, financial support, etc. Innovative strategies to reduce costs and increase revenues include implementing a volunteer program with local community groups or schools (as in Nabweru, Uganda) and sharing facilities with other organizations such as local government offices, schools, or community radio stations. Branching out to provide new services is another option—for example, the Timbuktu telecenter is also an ISP that can provide Internet access to patrons who decide to buy their own computers, and desktop publishing of wedding and funeral announcements has become a popular service in South African telecenters.

LINKS TO POLICY

Evaluators should bear in mind that data from pilot projects can be valuable for telecommunications policymakers and regulators. Interviews with officials in ministries of communications and regulatory bodies could help to identify key issues and information needs. Many countries are attempting to implement universal service or access policies. Evaluators could provide information that would show whether telecenters are in fact a viable means of providing rural and community access to advanced services such as the Internet. They could also provide the specific information needed for planning purposes such as traffic volumes, revenues from telecommunications services, and communities of interest. Some of these data may also be useful for oversight of the telecommunications sector. For example, data on line quality, outages, and time required to restore service can help regulators monitor the performance of licensed telecommunications operators.

DISSEMINATING RESULTS

The philosophy of evaluation as learning should also be reflected in how the information gained is shared with the various stakeholders. Simply sending them a copy of a research report may not be sufficient. The evaluators should be prepared to return to the community and to meet with the project staff and local stakeholders to explain the findings. They may also be able to help the staff to think about the changes that could be made in response to the feedback, such as providing outreach to underserved target groups, organizing more opportunities for training new users, extending the hours of operation, etc. Evaluators may also be able to suggest other opportunities for learning and sharing information such as workshops for managers and staff and exchange visits between similar projects.

The evaluation results should also be made available to other stake-holders, such as funding agencies, government ministries, and NGOs. Presentations at conferences, papers in journals, and postings on the Web can also help to disseminate the findings to others interested in ICT project evaluation around the world.

EVALUATING ICT PROJECTS FOR EDUCATION

ICTs are extremely valuable tools for providing access to a wide range of educational opportunities for young people and adults within and outside the formal education system. However, they pose significant challenges ranging from the need for relevant content, to the financial and human resources required to operate them, to the difficulty of integrating them within traditional curricula and institutions. Careful planning and evaluation can help to identify issues and strategies needed to gain maximum benefit from applications of ICTs in education.[2]

Evaluation of ICT projects in education can build on general ICT evaluation frameworks and draw lessons from approaches to evaluation suggested above for telecenters. However, the specific evaluation design must be tailored to education initiatives. In this section, evaluation issues and strategies for school networking projects (sometimes called schoolnets) are presented.

GOALS AND STAKEHOLDERS

To plan the evaluation, it is important to identify the objectives of introducing schoolnets, such as:

- To enable students to gain skills in using computers and services such as e-mail and the Internet;
- To prepare students for jobs requiring ICT skills;
- To improve the quality of education through:
  - helping teachers to develop and illustrate lessons;
  - enabling students to access supplementary course-related materials;
  - enabling students to exchange information or collaborate with distant students;
  - enable students to do research for class projects, etc.
- To extend access to ICTs to disadvantaged communities.

As with other projects, perceptions of goals may vary among the stake-holders, so that in addition to reviewing project documents, it is important to ascertain goals and expectations from stakeholders such as teachers, administrators, and the Ministry of Education. However, it is also important to gain an understanding of the expectations of the students themselves and, if possible, their parents. For example, students may see computer skills as the key to getting a job; parents may also be concerned about future employment prospects of their children, but may also worry about additional costs of ICTs or exposure of their children to inappropriate content. In Uganda, the PTA (Parent Teacher Association) played an active role in supporting ICTs in schools, and contributed to the schoolnet evaluation.[3]

To avoid a technology-driven mentality toward the project, stakeholders should be asked to identify problems that they hope the schoolnet project can help to solve. These may range from community issues such as high youth unemployment and lack of activities for young people, to the lack of educational materials in the schools to the need to motivate teachers or to reform the curriculum. Stakeholders should also be asked what they want to learn from the project, and how the information collected in the evaluation can be made useful to them. Once goals and expectations are identified, evaluators can "work backwards" to determine what conditions must exist in order for these goals to be achieved.

THE CHAIN OF INFERENCE FOR SCHOOL NETWORKS

The general premise of schoolnet projects is that access to computers that are networked for e-mail and Internet access can be beneficial for education. However, to achieve educational benefits requires many steps, including:

Access:
- The schools must be equipped with enough computers to enable students to use them;
- The students must have adequate access to the computers (e.g., one hour per week would be insufficient);
- The computers must be linked to a telecommunications network and ISP (Internet service provider);
- The equipment must work reliably, including computers, power supplies, telephone lines or other links, peripherals;
- The services must be affordable: the school must be pay for the Internet access, spare parts, and maintenance, etc.;

Skills:
- The students must learn the skills to use the software and network services;
- Teachers must also know how to use the equipment and services;

Applications:

- Software suitable for the curriculum must be available;
- The teachers must learn how to incorporate the information available through the facilities into the curriculum;
- The teachers must also be willing to incorporate the use of computers in the curriculum.

A break in any of these links can result in few significant benefits from the project. For example, if the computers break down frequently or the telephone line is unreliable, the schools will not be able to put these tools to use. If there are so few computers that students can use them for only an hour a week, or if the computers are locked up when the students are free to use them, there will be little benefit. If the schools cannot afford installation or access charges for telecommunications, the benefits of sharing information by e-mail and finding information on the Internet will not be achieved. Even if the facilities work reasonably well and are relatively accessible, if the teachers are not able to incorporate their use in the classroom and the curriculum, much of the project's potential benefit may not be realized (although motivated students who learn to use the computers may benefit to some extent).

MONITORING USAGE

As was true for telecenters, an important element of the evaluation is monitoring to determine how the school networks are being used, for example:

- Which teachers are using the school's network?
- How are these teachers using ICTs—for example, to prepare lessons or illustrate concepts?
- What percentage of students use the school network? On average, how much time per week do students use it?
- How are students using computers: for example, do they use computers to work on school subjects or to learn computer skills or to entertain themselves?
- Are teachers or students using e-mail? Internet access? If so, for what purposes? If not, why not?

If it is found that the schoolnets are not being used as anticipated, it will be important to identify problems such as equipment malfunctions, lack of training for teachers, difficulties in making computers available for students before or after school, lack of suitable content, and so forth.

LEARNING OUTCOMES

It is tempting to attempt to measure changes in performance on standard examinations or other tests that purport to measure learning. However, before developing a strategy to collection such information, it is important to determine how much access students have to computers, whether teachers have used the computer resources in preparing their lessons, and whether teachers have assigned projects requiring computer use that are related to the curriculum.

If students have had very little exposure to computers, the impact on learning may be minimal. If the computer resources have not been integrated into the curriculum for set subjects, students with computer access could not be expected to perform differently on set examinations. However, if teachers have attempted to foster independent research using the schoolnet, it would be important to develop a technique to compare students' abilities to seek answers and solve problems either before and after or with and without schoolnet access.

OTHER ISSUES FOR SCHOOLNET EVALUATION

Among other issues that schoolnet evaluations should address include:

• **Sustainability:** An analysis of costs associated with setting up, operating, and maintaining computer networks. This is important both to help the participating schools estimate the operating costs of sustaining their networks, and to provide planning information for other schools and educators intending to install computer networks with Internet access. This information could then be used to help the schools to prepare a schoolnet plan. (A similar planning exercise is required of U.S. schools applying for subsidies under the E-Rate program [see chapter 6]. This plan must be approved at the state level before they are eligible to receive the subsidy).[4]

• **Community access:** Evaluation can help to determine whether (or under what conditions) school networks can provide access to other parts of the community, and/or whether community access sites such as telecenters can be used to provide access for students and teachers.

• **Changes in attitudes and/or behavior:** Exposure to ICTs may influence student and teacher attitudes not only toward ICTs but toward education. For example, a WorldLinks evaluation of schoolnets in developing countries notes that some students participating in the project reported that they developed a more positive attitude toward school and improved their school attendance (Kozma et al., 1999).

EVALUATING TELEMEDICINE PROJECTS

Pilot projects and trials can help us to discover how information and communication technologies can contribute to better and more cost-effective health care. Evaluation of these projects can provide both feedback to participants on how well the projects are achieving their objectives and lessons to be shared among health care practitioners on how these tools can contribute to improving the quality and availability of health care in the developing world. Yet there is relatively little evaluative material on the effectiveness of telemedicine as a means of providing care—both in terms of clinical effectiveness and cost effectiveness.

One explanation is that most telemedicine projects are fairly recent, and the current emphasis in medicine on outcomes research, including quality, appropriateness, and cost effectiveness of care is also recent (Witherspoon, Johnstone, & Wasem, 1993). However, models do exist from experimental satellite projects carried out in the United States and Canada in the 1970s and 1980s, and during the past decade by the European Union (Hudson, 1990, 1997). A very useful evaluation resource, although its examples are primarily from U.S. telemedicine projects, has been published by the Institute of Medicine of the U.S. National Academy of Sciences (Field, 1996). The ITU and WHO have also published papers and conference proceedings on telemedicine, with an emphasis on projects in developing regions.[5]

A research design that will allow comparison with and without or before and after the intervention is an important consideration for telemedicine evaluation. Various quasi-experimental designs may be appropriate, such as control groups, which could be achieved by including similar sites with and without telemedicine: rural clinics, district hospitals, regional hospitals, etc. An alternative is random assignment of the intervention, but medical priorities, access to telecommunications, and/or political pressures often make random assignment of ICT interventions impractical.

TELEMEDICINE EVALUATION PLANNING

The same frameworks and approaches to evaluation presented earlier may be adapted for telemedicine project evaluation. Questions to be considered in telemedicine evaluation planning include:

- What are the major health problems (causes of mortality and morbidity) in the region covered by the project?
- What are the major challenges facing the health care system nationally and in the region covered by the project?

- What are the goals of the project? How do these goals relate to high priority health problems and challenges facing the health care system?
- Who are the major stakeholders? For example: patients and clients of the health care system, Ministry of Health, funding agencies, university medical faculty, academic researchers, health care staff at various levels: rural clinics, community health centers, district hospitals, regional hospitals, national medical centers, etc.
- What is the organizational structure of the health care system? For example, what is the chain of command for medical and administrative supervision; what is the hierarchy for patient referrals?
- What infrastructure and facilities are available at the various sites involved in the project (e.g., power supply, telecommunications services and reliability, medical equipment)?
- What are the capital costs of the project? Who is paying for these costs?
- What are the estimated operating costs: in terms of personnel, facilities, communications charges, etc.? How will these costs be covered during the project period? Is there a source of funding to continue after the pilot phase?

With the above inputs, it should be possible to identify the questions to be addressed by the evaluation. These are likely to include:[6]

- **Quality of Care:** What were the effects of the project on the process of health care? Are patients being treated sooner? With greater expertise? Closer to home rather than in distant centers?

- **Access to Care:** Does the use of telemedicine extend access to scarce medical resources? For example, can city-based specialists make their expertise available through consultations with rural colleagues? Or are there other factors that hinder the success of telemedicine facilities for these goals?

- **Outcomes:** What were the effects of the project on health outcomes—immediate, short term, and long term? Saving lives is the most dramatic short-term example—this may occur if emergency help is provided sooner than before (e.g., calling an ambulance or paramedic), if life threatening conditions are identified sooner (such as complications during delivery), or if attending health staff can get timely advice for critical cases.

  Intermediate term effects may be related to identification of chronic health care problems that would otherwise not have been treated, better monitoring of chronically ill patients, effective community health strategies.

Long-term effects may be measurable in terms of mortality and morbidity, but difficulties in isolating causality and tracking a large enough number of cases over a long enough period are likely to make such conclusions difficult to reach. Indicators that point to longer term trends may be all that is possible to measure during the life of the project, for example, number of cases treated with intervention of telemedicine, and disposition of these cases compared to all other cases of this illness or injury.

It may be possible to track the effect of telemedicine on case management. For example, was the original diagnosis confirmed or changed? If it was changed, what was the effect? Some of this information may be included on patient record forms, so that health care staff can code a few items on use of telemedicine in addition to other information routinely recorded. Information on outcomes may then be gathered by tracking the medical histories of patients treated with telemedicine facilities, or by interviewing health care providers about outcomes of such cases.

- **Relationship to Priorities:** How do the outcomes and other effects of the project relate to health problems and priorities of the health care system? Does the project have a positive impact on major health problems? Does the project help to address a priority of the health care system such as upgrading skills of rural health care providers, or help to solve a problem such as extending access to scarce expertise? This assessment will likely be critical in determining whether the project should be continued or expanded.

Cost Issues and Cost Effectiveness

As with other ICT projects, it will be important to identify project costs such as equipment, communication charges, personnel, training, technical support, etc. These costs should then be analyzed to estimate the ongoing costs of operating the project after the pilot phase, and the cost of extending or replicating the project.

The evaluation should also attempt to quantify cost savings and other benefits. For example, there may be savings if the health care system does not have to evacuate a patient to the city, or can discharge a patient sooner, with follow-up care being given in the home community. Early diagnosis and treatment can reduce costs of hospitalization and medication.

Some cost savings may be considered externalities (i.e., not directly attributable to the health care system). For example, patients who can be treated locally may save expenses of transport and accommodation required if they had to go to the city for treatment; patients treated early or more effectively may lose fewer days from work, benefiting their families and the general economy. Of course, other benefits such as saving lives may be very significant but not quantifiable.

Attitudes of Participants and Stakeholders:

Information should be collected on attitudes of all participants ranging from specialists to district physicians and community health workers, as well as patients. Issues to address could include:

- Patient concerns about privacy and confidentiality;
- Patient understanding of how and why telemedicine facilities were being used;
- Patient and practitioner confidence in remote diagnosis or advice;
- Estimates of time and effort required to use equipment;
- Fit within organizational structure and culture of the health care system;
- Demands on scarce resources: space, staff time, phone lines, other equipment, etc.;
- Perceived value for treating different types of medical problems.

Information on such attitudes and perceptions can be gathered through questionnaires designed for participating medical staff, interviews with patients, and/or focus groups.

As with evaluation of other ICT projects, findings from evaluation should be fed back to the participants as well as to other stakeholders to enable them to make any recommended adjustments in the project during its pilot phase and to consider the evaluation results in deciding whether to continue or expand the project. However, both evaluators and stakeholders should keep in mind that:

> widespread adoption of effective telemedicine applications depends on a complex, broadly distributed technical and human infrastructure that is only partly in place and is being profoundly affected by rapid changes in health care, information, and communications systems. The difficulties encountered during more than two decades of work to implement integrated information systems suggests the importance of persistence and realism for those working to demonstrate telemedicine's promise. (Field, 1996, p. 207)

Measuring Outputs and Outcomes of ICT Projects

The following are indicators proposed by the Appalachian Regional Commission for evaluation of community ICT projects:

Output Measures:
- Improved community access to telecommunications and information technology;
- Creation of a community access facility with advanced telecom services and staffing with technical support, available to all residents;
- Number of participants, residents, trainees, workers, teachers, etc. served by project;
- Number of residents students, trainees, workers, etc. who completed the programs;
- Number of residents, students, trainees, workers, etc. who have used equipment procured by project;
- Number of businesses served by project;
- Procurement and installation of new facilities, telecom equipment, and computers;
- Number of telecom sites established;
- Implementation of new curricula that meet higher quality educational or industry instructional standards;
- Number of technical or instructional staff hired to carry out project.

Outcome measures:
- Estimate or attribution of project's effects on measured student or trainee achievement;
- Number of students or trainees who completed a course that meets recognized academic and industry standards;
- Number of professionals, students, or trainees who obtained a relevant certification or recertification;
- Number of participants receiving professional development services and expected benefits;
- Number of professionals, students, trainees, or residents who have developed technically related skills or enhanced technical literacy;
- Number of businesses that have adopted telecom-based e-business practices.

Project implementation:
- Existence of community or regional strategic plan that led to project implementation;
- Determination of whether project was on schedule and on budget;

- Evidence of project monitoring and evaluation throughout project cycle;
- Sustainability of project following initial startup funding;
- Post project uses of equipment/facilities procured during funded project.[7]

NOTES

1. See www.infodev.org
2. Additional resources useful for planning and evaluating ICT projects in education include: Hudson (1998, 1992); Scharffenberger (1999); CSIR (1999); SRI International (1998); Latchem and Walker (2001).
3. "World Case Study Reports: Uganda." Washington, DC: World Banks Links for Development Program, 1999.
4. Universal Service Administrative Company: Schools and Libraries. http://www.sl.universalservice.org
5, See, for example, "ITU-D Technologies Infrastructure and Applications: Telemedicine" www.itu.int/ITU-D/tech/telemedicine/ and Epstein and Vernaci (1998).
6. Adapted from Field (1996, pp. 11–15).
7. "Request for Proposals from The Appalachian Regional Commission for a Program Evaluation of ARC's Telecommunications projects, June 1, 2001. www.arc.gov

Chapter 10

From Global to Local

We have now reached the stage when virtually anything we want to do in the field of communications is possible. The constraints are no longer technical, but economic, legal or political.

—Arthur C. Clarke (1992, p. 224)

This book has explored how ICTs can contribute to rural development, and what steps are needed to extend affordable access and facilitate effective use. This final chapter examines some of the issues still to be addressed.

BEYOND TECHNOLOGY

Each new communication technology has been heralded as offering numerous benefits. Satellites and cable television were to provide courses taught by the best instructors to students in schools, homes, and workplaces. Videoconferencing was to largely eliminate business travel. Telemedicine was to replace referral of patients to specialists. Computers were to replace traditional teaching with more personalized and interactive instruction.

To some extent all of the prophesies have been fulfilled, yet the potential of the technologies is far from fully realized. In many cases, it took institutional change and incentives to innovate for these technologies to have much effect. In North America, the more remarkable change is in these incentives rather than the technologies. As school districts face shrinking budgets and new curricular requirements, as spiraling health care budgets

are targeted by governments and insurance companies, and as business re-
alizes that people must "work smarter" to compete in a global economy,
they find new and compelling reasons to turn to telecommunications and
information technologies.

Thus, investment in technology alone will not likely result in major social
benefits. Policymakers and development agencies appear aware that public
sector investment is needed to foster new educational and social service applications; there is widespread belief in the need to fund trials and demonstration projects. Yet seed money for pilot projects may not ensure long-
term implementation. Schools with Internet access will benefit if the
available content can help to address their educational priorities. If the services are perceived as frills, or if there is no budget allocation to buy computers or pay monthly usage charges, connection to the Internet will mean
little. Similarly, if insurers will not authorize payment for teleconsultations,
or physicians are not authorized to practice beyond their borders, telemedical applications will remain limited. And if prices for connection and
usage are beyond the reach of low income and rural residents, small businesses, and nonprofit organizations, the much-heralded information society will be very narrowly based.

THE BOTTOM OF THE PYRAMID

As noted in the earlier chapters on policy, telecommunications operators
often assume that demand is too low and/or costs are too high to provide
service for low-income people, particularly in rural areas. However, as C. K.
Prahalad (2005) points out, with the right strategies, businesses can profit-
ably serve poor people and enable them to improve their own economic
conditions. The growth of wireless phones in the developing world shows
that demand can be more than sufficient if pricing is affordable. Prepaid
phone service has been the primary mechanism for enabling customers to
"pay as you go" and to eliminate risk for service providers. This is just one
example of how telecommunications services can be made available to poor
people near the bottom of the economic pyramid.

Individual entrepreneurs, primarily women, now make a living selling
prepaid phone cards in Uganda, Mozambique, and many other poor countries. In Bangladesh, village women operate "portable payphones," selling
individual calls on cellular phones that they carry with them. In the Philippines, entrepreneurs may even resell small increments of talk time by transferring units from one cellphone to another, to provide access to people to
make a single call if they cannot afford to buy a whole prepaid card. The
mobile phone operator has developed other applications for this technology, so that it can also be used to transfer remittances. A Filipina maid in

Hong Kong may transfer funds directly to a relative in the Philippines by visiting a Hong Kong vendor with an international cellphone. She pays the vendor, who can immediate transfer the "electronic cash" to the recipient's cellphone. The recipient can then exchange the electronic cash for a stored value cash card or for Philippine pesos at a local shop.[1]

ECONOMIC VERSUS SOCIAL GOALS: "WHEN THE DOOR OPENS . . ."

A paradox not fully acknowledged by some developing country leaders is that the inevitable result of investing in information infrastructure is to increase access to information. Telecommunications planners and policy-makers in both developing and industrialized countries must recognize that the sharing and utilization of information, and not the mere extension of networks, should be the ultimate purpose of telecommunications policy reform.

Some governments that see information technology as critical to their economic development strategy are at the same time concerned about the sociopolitical implications of access. One of the most ironic examples of the simultaneously held goals of modernization and control is Singapore, which staked its economic future on becoming an "intelligent island." Yet Singapore has retained tight control over individual access to information. The government applies broadcast content regulations to the Internet, holding Internet service providers accountable for content accessible to their customers. Also, it is illegal for individuals to install satellite antennas, so that Singaporeans cannot watch satellite-transmitted programs from regional satellites, including those uplinked from Singapore's own high tech industrial parks.

Although credited with introducing market reforms in the Chinese economy, Deng Xiaoping voiced his ambivalence about opening China's doors to the world: "When the door opens, some flies are bound to come in" (quoted in Schwankert, 1995, p. 112). China may be the world's largest market for telecommunications with the fastest growing number of mobile phone and Internet users, but the government is reluctant to allow access to information from abroad. Government attempts to control access include banning satellite antennas, blocking access to Internet sites, and impeding access to the Internet itself and to other means of electronic communication. The Chinese government has also attempted to censor text messages, with the ostensible rationale of filtering out pornographic and fraudulent messages; however, political content is considered the real target. Chinese mobile phone users sent more than 220 billion text messages in 2003, more than half the world's total (Lim, 2004).

Malaysia's Multimedia Super Corridor (MSC), with its optical fiber networks and multimillion dollar test bed, links Kuala Lumpur with the new administrative capital, Putrajaya, and its new international airport. The government provides tax concessions to approved multimedia companies and allows foreign firms to bring in skilled foreign workers. Yet Malaysia faces the problem of "reconciling Western ideas in a traditional, mostly Islamic society with pervasive censorship" (Langdale, 1998). The government's proposed compromise is to permit open access to the Internet within the Multimedia Supercorridor, but retain censorship in the rest of the country. The government has also pledged to protect intellectual property rights within the MSC; however, pirated software is still readily available in major cities.

Clever users will inevitably find means to bypass these roadblocks, as shown by dissidents' use of facsimile and electronic mail during the Tiananmen Square uprising in China, the proliferation of satellite antennas in countries where they are officially banned, and the widespread availability of supposedly illegal services that undercut international tariffs such as callback and voice-over-Internet telephony. Yet these strategies are likely to be limited to an elite few with the technical knowhow or political connections to end-run the regulations. Only where governments recognize that suffocation is worse than a few flies will the gaps really disappear.

UNDERSTANDING TRENDS IN ACCESS

Despite efforts to extend access to ICTs, we still need to learn which underlying factors are the best explanations of variations in access. For example, in the United States, attention is frequently focused on ethnicity and rurality (e.g., access by blacks, Hispanics, native Americans; disparities between urban and rural residents). However, other factors such as income and education (often highly correlated) appear to be more influential. Similar analysis in other countries may reveal underlying factors that form barriers to access. As definitions of access change to include computers and Internet connectivity as well as telephone service, we could expect that education would be an increasingly important predictor of access, since better educated people would tend to have better technical skills and greater perceived need for use of computers and online services.

The ITU has proposed a Digital Access Index (DAI) to measure "overall ability of individuals in a country to access and use ICTs." It identifies five fundamental variables: availability of infrastructure, affordability of access, educational level, quality of ICT services, and Internet usage. "If the infrastructure is not available, there can be no access. If the population cannot afford to pay for ICT products and services, there can be no access. If citizens do not have a certain level of education, they will not be able to use newer ICTs such as computers or the Internet. If the ICT experience is poor, people will either cease using them or be incapable of using them effectively or creatively."[2]

Beyond access, it will be important to understand what factors influence use of information services once they are accessible, either through individual ownership and connectivity or public sites such as schools and libraries. Are there other factors such as computer use at school that are likely to encourage access? Are there strategies such as community access or training that could increase utilization? Among youth, are there specific factors such as exposure at an early age, that appear preconditions for later use? Among adults, are there information-seeking behaviors or social norms that may influence use of ICTs? In some cultures, women may be discouraged from using technology; also, older or less educated people may feel more comfortable using an infomediary such as a librarian, social worker, or extension agent to find information they need or to contact others electronically. For example, peasant farmers in Ecuador found out how to eliminate a pest that was destroying their potato crop through the assistance of a field worker who posted their question on several Internet news groups (personal communication, October 1999).[3]

NO RECIPES FOR INSTANT DEVELOPMENT

It is tempting to see ICTs as a "quick fix" for development. As noted in Chapter 9, a danger of the recent enthusiasm about ICTs is creation of unrealistic expectations among policymakers and donors that investing in infrastructure and funding ICT projects will lead to significant immediate developmental impacts. While there may be short-term benefits when critical advice can save lives or crops, timely market information can help producers to get higher prices, and online tutorials can provide new opportunities for learning, the impact of these and many other uses of ICTs is likely to take more time and be difficult to measure. In addition to devising models and conducting field evaluations, researchers can help by explaining the contexts in which ICTs are introduced and the logical chain of inference between ICT interventions and development incomes.

To make a contribution to our understanding of how new technologies can contribute to social, economic, and cultural development goals, ICT projects must at minimum:

- Be designed to contribute to priority goals, rather than being driven exclusively by technology;
- Take into consideration the local environment in terms of availability of equipment, skills, and other resources;

- Involve the users;
- Include external evaluation;
- Plan for sustainability past the project period.

Although policymakers may be aware that public sector stimulus is needed to foster applications of new technologies, support for trials and pilot projects may not ensure long-term implementation. If the services are perceived as frills diverting energy and resources from higher priorities, or if there is no plan to sustain operations past the pilot phase, project funding will have bought only an expensive distraction.

THE PUBLIC INTEREST

A public interest perspective on connectivity can focus attention on the importance of ICTs in development, and act as a lens through which to view progress toward connectivity for all. This approach assumes a broadening of the definition of "public interest" beyond the assessment of connection to the network. It involves an analysis of the potential benefits of access to education and social services, the impact of geographical as well as income-related disparities, and the potential economic benefits of affordable access to information for both individual and collective activities.

This approach provides a perspective to remind us that investment in technology alone will not likely result in major social, economic, cultural benefits. As noted in earlier chapters, in addition to extending networks, there are also likely to be ongoing policy issues such as the need to ensure that quality of service meets specified targets, that pricing is affordable, and that benchmarks are reviewed and revised when necessary.

However, it is important to note that more than access will likely be necessary to achieve significant benefits. Effective applications of these facilities may also require training, mentoring, and, in certain cases, facilitation through intermediaries. A workforce with sufficient general education and specialized training as well as an institutional environment that fosters innovation and productivity are likely to be critical factors. Sustainability will continue to be a concern, particularly in maintaining noncommercial forms of access such as through schools and libraries, and nonprofit community access centers.

This book has shown how innovative technologies, strategies, and policies can help to increase access to communication services, and in turn, contribute to development. The operative word here is *can*. Whether they will depends on the continued commitment of the many stakeholders, and an unwaivering focus on the needs of the users.

NOTES

1. See www.smart.com.ph/smart
2. See International Telecommunication Union (2003).
3. Heather E. Hudson, field research and unpublished reports, 1999.

A Selection of Useful Web Sites on ICTs and Development

African Information Society Initiative: www.uneca.org/aisi/

African Virtual University: www.avu.org

Alaskan Federal Health Care Access Network (AFHCAN): www.afhcan.org

Association for Progressive Communications (APC): an international network of civil society organizations concerned with using ICTs for development: www.apc.org

Bridges: www.bridges.org: an organization concerned with using ICTs for development.

Cisco Networking Academies: www.cisco.netacad.net/public/academy/About.html

Columbia University Institute for Tele-Information: www.citi.columbia.edu (See Events, Past Events, Publications)

Commonwealth Telecommunications Organization: www.cto.int

Communication Technology Centers' Network: a U.S.-based organization concerned with using technology for community development: www.ctcnet.org

Development Gateway: a portal that provides information and links to resources on development: http://topics.developmentgateway.org/ict

Digital Divide Network: an Internet community for educators, activists, policymakers, and concerned citizens working to bridge the digital divide: www.digitaldividenetwork.org

DOT-COM Alliance: a group of USAID-funded ICT initiatives for developing countries: www.dot-com-alliance.org

E-Inclusion: an initiative of Hewlett-Packard (HP): www.hp.com/e-inclusion/en/

Federal Communications Commission (U.S.): www.fcc.gov

Global Knowledge Partnership: an organization providing information and resources on ICTs for development: www.globalknowledge.org

HealthNet and SatelLife: www.healthnet.org

158

Information for Development Program (InfoDev): an international consortium based at the World Bank that supports developing country initiatives in ITC applications and policies: www.infodev.org

International Development Research Centre: a Canadian organization with a major focus on ICTs for development: www.idrc.ca (See Evaluation, Books Online, ICT4D)

International Telecommunication Union, Telecommunications Development Bureau: www.itu.int/ITU-D/ict/

Keewaytinook Okmakanak (KNET): a Canadian indigenous organization using ICTs for education, health, and community development: www.knet.ca

Learnlink: materials on ICT projects in 17 developing countries: http://learnlink.aed.org

LINCOS: Little Intelligent Communities: www.lincos.net

LIRNE.NET: Learning Initiatives on Reform for Network Economies, a consortium concerned with facilitating ICT-related institutional reform and regulatory training: www.lirne.net

National Telecommunications and Information Adminstration (U.S.): www.ntia.doc.gov

Organization for Economic Cooperation and Development (OECD): www.oecd.org (See Information and Communication Technologies)

Rural Utilities Service (U.S.) ICT projects: www.usda.gov/rus/telecom.index.htm

SANGONeT: an organization based in South Africa, concerned with linking civil society through ICTs: www.sangonet.org.za

SchoolAccess: a suite of services designed for schools in Alaska and other rural and underserved areas of the U.S.: www.schoolaccess.net

Schoolnet Africa: an organization concerned with providing ICT facilities, training, and applications for African schools: www.schoolnetafrica.net

Smart Communities (Canada): a resource on broadband and other ICT projects for Canadian communities: http://smartcommunities.ic.gc.ca

Telecommunications Act of 1996 (U.S.): see www.fcc.gov/telecom.html

Telecenters: see www.telecentre.org (site primarily for telecenter practitioners; information on telecenters is available also on several other sites)

United Nations ICT Task Force: www.unicttaskforce.org

Universal Services Administrative Company (U.S.): responsible for administering funds for Internet access for schools, libraries and rural health centers: www.universalservice.org

World Bank, Global Information and Communication Technologies Department: http://info.worldbank.org/ict/

World Dialogue on Regulation for Network Economies: an organization concerned with improving regulation and governance for network economies: www.regulateonline.org

World-Links: an organization concerned with improving education and development opportunities through use of ICTs for learning: www.world-links.org

Glossary

Acacia: The Acacia Initiative on Communities and the Information Society in Africa, a program sponsored by the International Development Research Centre (IDRC), to empower sub-Saharan communities with the ability to apply ICTs for their own development.

AFHCAN: Alaska Federal Health Care Access Network.

Analog: A signal that varies continuously (as contrasted with a digital signal).

APEC: Asia Pacific Economic Conference, an association of countries in Asia and on the Pacific Rim.

ASEAN: Association of Southeast Asian Nations.

ATM: Asynchronous Transfer Mode, a high-speed switching technique that uses fixed-size cells to transmit voice, data, and video.

AT&T: American Telephone and Telegraph.

ATS: Applied Technology Satellite, a series of U.S. experimental satellites (e.g. ATS-1, ATS-6).

Bandwidth: The capacity of a communications channel, expressed in hertz (cycles per second).

Bit: Binary digit, i.e. either 0 or 1.

Broadband: High speed, high capacity transmission, typically 384 kbps or greater.

Bypass: Telecommunications transmissions that avoid some or all of the public communications network.

CAT-scan: Computerized axial tomography scan, an x-ray procedure which combines many x-ray images with the aid of a computer to generate cross-sectional views of the internal organs and structures of the body.

C-band: Portion of the electromagnetic spectrum in the 4-to-6 GHz range, used for satellite communications.

CDMA: code division multiple access, a transmission technique used in digital wireless technology.

Cellular communications: A wireless communications system in which the coverage area is divided into sections called cells. A transmitter/receiver in each cell relays calls to wireless telephones and devices within each cell, and passes traffic from one cell to another.

CIRCIT: Centre for International Research on Communication and Information Technologies (Australia).

COL: Commonwealth of Learning, an intergovernmental organization established by Commonwealth heads of government, focusing on open learning and distance education.

Common carrier: An entity that provides telecommunications transmission services to the public.

Compression: The process of eliminating redundant information from a data stream before it is stored or transmitted.

CRTC: Canadian Radio-television and Telecommunications Commission.

CTO: Commonwealth Telecommunications Organization.

CTS: Communications Technology Satellite, an experimental satellite that was jointly operated by the United States and Canada.

DAI: Digital Access Index, a set of measures to compare ICT access across nations developed by the ITU.

DBS: Direct broadcasting satellite, designed to transmit to very small terminals for household television reception.

Digital: A discrete or discontinuous signal that transmits audio, data, and video as bits (binary digits) of information.

Digital switching: A process in which connections are established by processing digital signals without converting them to analog signals.

Downlink: An earth station used to receive signals from a satellite; the signal transmitted from the satellite to earth.

DSL: Digital subscriber line, a technology that allows voice and high-speed data to be transmitted over a single telephone line.

DTH: Direct-to-home, referring to satellites designed to transmit video to very small satellite antennas.

E-Rate: Education rate, referring to subsidies for Internet access for schools and libraries, part of the U.S. Universal Service Fund.

Earth station: The antenna and associated equipment used to transmit and/or receive signals via satellite.

EKG or ECG: Electrocardiogram.

Encryption: Transformation of data for transmission so that it cannot be understood if intercepted.

FAO: Food and Agricultural Organization of the United Nations.

FCC: Federal Communications Commission (U.S.).

Frequency: Cycles per second (expressed in hertz).

Frequency Spectrum: The range of frequencies useful for radio communications, from about 10 KHz to 3,000 GHz.

G7: A grouping of the world's seven major industrialized nations: Canada, France, Germany, Japan, Italy, the United Kingdom, and the United States.

G8: The G7 countries plus Russia.

Gbps: Gigabits (billions of bits) per second.

GDP: Gross Domestic Product.

GEO: Geostationary satellite, a satellite with an equatorial orbit 22,300 miles (36,000 kilometers) above the earth. The satellite revolves around the earth in 24 hours, and thus appears stationary when viewed from earth.

GHz: Gigahertz; billions of hertz (cycles per second).

GII: Global information infrastructure.

GNP: Gross National Product.

GPRS: General Packet Radio Service: a standard that allows higher rates of transmission (up to 115 kbps) over GSM networks.

GPS: Global Positioning System.

GSM: Global System for Mobile Communications, a European digital cellular standard, widely used around the world.

GSO: Geostationary or geosynchronous orbit. See GEO.

HF: High frequency; the frequency band from 3 to 30 MHz (also known as shortwave).

HFC: Hybrid fiber-coax, a combination of optical fiber and coaxial cable used to deliver broadband services to end users.

ICT: Information and communication technologies.

ICT4D: Information and communication technologies for development.

Intelsat: The International Telecommunications Satellite Organization, now privatized but originally a consortium of national telecommunications organizations, that provides global satellite services.

IEEE: Institute of Electrical and Electronic Engineers, a technical organization that has developed numerous communications standards.

Inuit: Indigenous people of the Canadian Arctic (known formerly as Eskimos); in Alaska, known as Inupiat and Yupik.

ISDN: Integrated Services Digital Network, a set of standards for public digital telecommunications networks.

ISP: Internet service provider.

ITFS: Instructional Television Fixed Service; microwave frequencies allocated by the FCC for educational use.

ITU: International Telecommunication Union, the United Nations agency responsible for telecommunications.

Ka-band: Portion of the electromagnetic spectrum in the 20-to-30 GHz range, used for satellite communications.

Kbps: Kilobits (thousands of bits) per second.

KHz: Kilohertz; thousands of hertz (cycles per second).

Ku-band: Portion of the electromagnetic spectrum in the 12-to-14 GHz range, used for satellite communications.

LAN: local area network; a network linking computers at a single location.

LDCs: Least developed countries.

Leland Initiative: A project sponsored by USAID to extend Internet connectivity in Africa.

LEO: Low earth orbiting satellite. In contrast with GEOs, LEOs appear to move across the sky, and must be tracked. A series of satellites is required to provide continuous coverage of a given geographical area.

LINCOS: Little intelligent communities, typically telecenters housed in modified shipping containers.

Local loop: The link between the subscriber's premises and the local exchange.

Mbps: megabits (millions of bits) per second.

MHz: Millions of hertz (cycles per second).

Microwave: Radio transmissions in the 4 to 28 GHz range, requiring antennas located within line-of-sight.

Modem: Modulator/demodulator; a device that enables digital data to be transmitted over analog circuits.

MMDS: Microwave multipoint distribution system, sometimes referred to as "wireless cable."

MRI: Magnetic resonance imaging, a noninvasive method of using a magnetic field and radio waves to produce detailed images of the inside of the human body.

Narrowband channel: A communications channel that transmits voice or data at relatively low data rates using limited bandwidth.

NECA: National Exchange Carriers Association (U.S.).

NII: National information infrastructure.

NGO: Nongovernmental organization.

NTIA: National Telecommunications and Information Administration, in the U.S. Department of Commerce.

OECD: Organization for Economic Cooperation and Development, an international organization of 30 member nations with generally democratic market-based economies.

Nunavut: Semi-autonomous Inuit territory in Arctic Canada.

Optical fiber: A communications medium made of hair-thin glass that conducts light waves.

PCO: Public call office, a public or pay telephone.

PCS: Personal Communications System or Service, cellular service using small handheld wireless phones or PDAs.

PDA: personal digital assistant, a handheld electronic device for storing addresses, calendar, etc.; may also be equipped to transmit and receive data.

Photovoltaic: Material that develops voltage and electrical current when light shines on it; used for solar power systems.

Point-to-multipoint service: Transmission of a signal from one originating point to many receiving points; broadcasting.

PSTN: Public switched telephone network.

PTT: Ministry of Posts, Telephone and Telegraph; a government organization that provides telecommunications services as well as postal services.

PUC: Public utilities commission, state-level regulator in the United States.

RUS: Rural Utilities Service, an agency in the U.S. Department of Agriculture (formerly Rural Electrification Administration).

STD: Standard trunk dialing, more commonly known as long distance dialing.

TCP/IP: transmission control protocol/internet protocol; two of the standard Internet protocols for data transmission.

TDMA: Time division multiple access; a multiplexing method that divides a circuit's capacity into time slots.

Teledensity: Main telephone lines per 100 population.

Telemedicine: Use of telecommunications for medical diagnosis, patient care, and health education.

TEMIC: The Telecommunications Executive Management Institute of Canada.

UNDP: United Nations Development Program.

UNESCO: United Nations Educational, Scientific and Cultural Organization.

UNITAR: United Nations Institute for Training and Research.

USAC: Universal Services Administrative Company (U.S.).

USAID: United States Agency for International Development.

USF: Universal service fund.

USO: Universal service obligation, typically a requirement that a telecommunications carrier provide service to all customers in a specified region, or to specified groups such as low income or disabled residents, often at concessionary rates.

USP: University of the South Pacific.

USTTI: United States Telecommunications Training Institute.

UWI: University of the West Indies.

VSAT: Very small aperture terminal, a small satellite dish.

WAN: wide area network; a computer network covering a large geographical area.

WHO: World Health Organization, the United Nations specialized agency for health.

WiFi: Wireless Fidelity, a term used for a set of wireless standards for local coverage, also known as 802.11 (established by the IEEE).

WiMAX: wireless broadband standard that does not require line of sight, also known as 802.16 (established by the IEEE).

WLL: Wireless local loop, using radio technology rather than twisted copper pair to reach subscribers' premises.

References

Appalachian Regional Commission. (2001). *Request for Proposals from the Appalachian Regional Commission for A Program Evaluation of ARC's Telecommunications Projects.* Washington, DC: Appalachian Regional Commission.

Atkins, G. S. (1990). Farm Radio in Developing Countries; A Case Study of the Developing Countries Farm Radio Network. *Development, 2,* 108–112.

Block, C. B. (1985). Satellite Linkages and Rural Development. In H. E. Hudson (Ed.), *New Directions in Satellite Communications: Challenges for North and South.* Norwood, MA: Artech.

Cairncross, F. (2001). *The Death of Distance: How the Communications Revolution is Changing our Lives.* Boston: Harvard University Business School Press.

Carlson, S., & McGhee, R. (2002). "World Links . . . Opening a World of Learning." June 19, 2002. http://world-links.org/english/assets/NECC2002.ppt

Carlsson, Sylvia. "Telemedicine in the Bush: A Glimpse into the New Century." www.telemedicine.alaska.edu

Cherry, C. (1957). *On Human Communication: A Review, a Survey, and a Criticism.* Cambridge, MA: MIT Press.

Cherry, C. (1971). *World Communication: Threat or Promise? A Socio-Technical Approach.* London and New York: Wiley-InterScience.

Chu, G. C., Srivisal, A. C., & Supadhiloke, B. (1985). Rural Telecommunications in Indonesia and Thailand. *Telecommunications Policy,* June, pp. 159–169.

Clarke, A. C. (1992). *How the World was Won.* New York: Bantam Books.

Crandall, R. W., & Waverman, L. (2000). *Who Pays for Universal Service?* Washington, DC: Brookings Institution.

Creating a Digital Dynamic: Final Report of the Digital Opportunity Initiative. (2001). New York: Markle Foundation.

Cronin, F. J., Colleran, E. K., Herbert, P. L., & Lewitzky, S. (1993). Telecommunications and Growth: The Contribution of Telecommunications Infrastructure Investment to Aggregate and Sectoral Productivity. *Telecommunications Policy, 17*(9), 677–690.

Cronin, F. J., Gold, M. A., Mace, B. B., & Sigalos, J. I. (1994). Telecommunications and Cost Savings in Educational Services. *Information, Economics and Policy, 6*(1), 53–75.

165

Cronin, F. J., Parker, E. B., Colleran, E. K., & Gold, M. A. (1991). Telecommunications Infrastructure and Economic Growth: An Analysis of Causality. *Telecommunications Policy, 15*(6), 529–535.

Cronin, F. J., Parker, E. B., Colleran, E. K., & Gold M. A. (1993). Telecommunications Infrastructure and Economic Development. *Telecommunications Policy, 17*(6), 415–430.

Defining Migration Policies in an Interdependent World. (2003). *IOM Migration Policy Issues,* No. 1, March.

Edgar, C. (2001). Outlook Never Brighter. *Communications Africa,* October/November, p. 2.

Epstein, D., & Vernaci, R. L. (1998). Telemedicine meets the Global Village. *Perspectives in*

Health, 3(2).

Evans, J. R., Hall, K. L., & Warford, J. (1988). Shattuck Lecture—Health Care in the Developing World: Problems of Scarcity and Choice. *New England Journal of Medicine, 305,* 1117–1127.

Field, M. J. (Ed.). (1996). *Telemedicine: A Guide to Assessing Telecommunications in Health Care.* Washington, DC: National Academies Press.

Fryer, M., Burns, S., & Hudson, H. (1985). Two-Way Radio for Rural Health Care Delivery. *Development Communication Report,* Autumn, pp. 5, 16.

Garretson, G. (2002) "Broadband group wants unified right of way plan." *PCWorld,* May 30. http://www.pcworld.com/news/article/0,aid,101550,00.asp

Garrett, L. (1994). *The coming plague: Newly emerging diseases in a world out of balance.* New York: Farrar, Straus, and Giroux.

Gidaspov, A. (2004). Trends in the Russian Telecoms Market. *U.S. Commercial Service,* March 1, p. 2.

Goldschmidt, D., Forsythe, V., & Hudson, H. (1980). *An Evaluation of the Medex Guyana Two-Way Radio Project.* Washington, DC: Academy for Educational Development, December.

Golladay, F. (1980). *Health Problems and Policies in Developing Countries.* World Bank Staff Working Paper, No. 402, Washington, DC, August.

Gore, A. Jr. (1991). Infrastructure for the Global Village: Computers, Networks and Public Policy. *Scientific American, 265*(3), p. 150.

Hafkin, N., & Taggert, N. (2001). *Gender, Information Technologies, and Developing Countries: An Analytical Study.* Washington, DC: Academy for Educational Development.

Hansen, S. et al. (1990). Telecommunications in Rural Europe: Economic Implications. *Telecommunications Policy,* June, pp. 207–222.

Hardy, A. P. (1980). The Role of the Telephone in Economic Development. *Telecommunications Policy, 4*(4), December, pp. 278–286.

Hawthorne, P. (2002). "Positively Sesame Street." *Time Europe,* Sept. 22. http://www.time.com/time/europe/magazine/article/0,13005,901020930-353521,00.html

Higgins, C., Dunn, E., & Conrath, D. (1984). Telemedicine: An Historical Perspective. *Telecommunications Policy, 8*(4), December, pp. 307–313.

High-Level Group on the Information Society. (1994). *Europe and the Global Information Society: Recommendations to the European Council* (The Bangemann Report). Brussels: European Commission.

Hudson, H. E. (1984). *When Telephones Reach the Village: The Role of Telecommunications in Rural Development.* Norwood, NJ: Ablex.

Hudson, H. E. (1990). *Communication Satellites: Their Development and Impact.* New York: Free Press.

Hudson, H. E. (1992). *Applications of New Technologies in Distance Education: Telecommunications Policy Issues and Options.* Melbourne, Australia: Centre for International Research on Communication and Information Technologies (CIRCIT), July.

Hudson, H. E. (1997a). Converging Technologies and Changing Realities: Toward Universal Access to Telecommunications in the Developing World. In W. H. Melody (Ed.), *Telecom Reform: Principles, Policies, and Regulatory Practices.* Lyngby, Denmark: Technical University of Denmark.

Hudson, H. E. (1997b). *Global Connections: International Telecommunications Infrastructure and Policy.* New York: Wiley.

Hudson, H. E. (1999). Beyond Infrastructure: A Critical Assessment of GII Initiatives. In S. E. Gillett & I. Vogelsang (Eds.), *Competition, Regulation, and Convergence.* Mahwah, NJ: Lawrence Erlbaum Associates.

Hudson, H. E. (2001). "Telecenter evaluation: Issues and strategies." In C. Latchem & D. Walker (Eds.), *Telecentres: Case studies and key issues.* Vancouver, Canada: Commonwealth of Learning.

Hudson, H. E. (2004). Universal Access: What have we learned from the E-Rate? *Telecommunications Policy, 28,* issues 3–4.

Hudson, H. E., Goldschmidt, D., Hardy, A. P., & Parker, E. B. (1979). *The Role of Telecommunication in Socio-Economic Development.* Report prepared for the International Telecommunication Union by Keewatin Communications, Washington, DC.

Hudson, H. E., Hardy, A. P., & Parker, E. B. (1982). Impact of Telephone and Satellite Earth Station Installation on GDP. *Telecommunications Policy, 6*(4), December, pp. 300–307.

Hudson, H. E., & Parker, E. B. (1973). Medical Communication in Alaska by Satellite. *New England Journal of Medicine,* December.

Hudson, H. E., & Pittman, T. S. (1999). From Northern Village to Rural Village: Rural Communications in Alaska. *Pacific Telecommunications Review, 21*(2).

Hundt, R. E. (1997). *Giving Schools and Libraries the Keys to the Future.* Speech to the National School Boards Association, Washington, DC, January 27.

Hundt, R. E. (2000). *You Say You Want a Revolution: A Story of Information Age Politics.* New Haven: Yale University Press.

Information Highway Advisory Council. (1994). *Canada's Information Highway: Providing New Dimensions for Learning, Creativity, and Entrepreneurship.* Ottawa, Information Canada, November.

Information Infrastructure Task Force. (1993). *The National Information Infrastructure: Agenda for Action.* Washington, DC: U.S. Department of Commerce, National Telecommunications and Information Administration.

International Commission for Worldwide Telecommunications Development (The Maitland Commission). (1984). *The Missing Link.* Geneva: International Telecommunication Union, December.

International Telecommunications Union. (1986). *Information, Telecommunications, and Development.* Geneva: ITU.

International Telecommunication Union. (1998). *World Telecommunication Development Report 1998.* Geneva: ITU.

International Telecommunication Union. (1999). *Challenges to the Network: Internet for Development.* Geneva: ITU.

International Telecommunication Union. (2001). *World Telecommunication Development Report 2001.* Geneva: ITU.

International Telecommunication Union. (2003). *World Telecommunication Development Report 2003.* Geneva: ITU.

Isaacs, S., "ICTs in African Schools: The Experience so far." See www.schoolnetafrica.net

Jensen, M. (1999). "The Wireless Toolbox: A Guide to Using Low-Cost Radio Communication Systems for Telecommunication in Developing Countries—An African Perspective." http://web.idrc.ca/es/ev-10592-201-1-DO_TOPIC.html

Jensen, M., & Walker, D. (2001). Telecenter Technology. In C. Latchem & D. Walker (Eds.), *Telecentres: Case Studies and Key Issues.* Vancouver, Canada: Commonwealth of Learning.

Kaul, S. N. (1978). *Benefits of Rural Telecommunications in Developing Countries.* Paris: OECD.

Kayani, R., & Dymond, A. (1997). *Options for Rural Telecommunications Development.* Washington, DC: World Bank.

Kozma, R. et al. (1999). World Links for Development: Accomplishments and Challenges. *Monitoring and Evaluation Annual Report 1998–1999.* SRI International, Menlo Park, California.

Langdale, J. V. (1998, January). The National Information Infrastructure in the Asia-Pacific Region. *Proceedings of the Pacific Telecommunications Conference,* Honolulu.

Lankester, C., & Labelle, R. (1997, June). *The Sustainable Development Networking Programme (SDNP): 1992–1997.* Paper presented at the Global Knowledge Conference, Toronto.

Latchem, C., & Walker, D. (Eds.). (2001). *Telecentres: Case Studies and Key Issues.* Vancouver, Canada: Commonwealth of Learning.

Lebeau, M. (1990). *Towards Gender Sensitive Research in Telecommunications.* Vancouver: INTELECON Research and Consultancy.

Leff, N. H. (1984). Externalities, Information Cost, and Social Benefit-Cost Analysis for Economic Development: An Example for Telecommunications. *Economic Development and Cultural Change, 32,* 255–276.

Lim, L. (2004). "China to Censor Text Messages." www.BBC.com

MacQueen, K. "Surfing the World from the Frozen North." *Ottawa Citizen,* February 23, 1997.

www.schoolnet.ca/aboriginal/learningcircle/05/frozen-e.html

Martin, W. J., & McKeown, S. F. (1993). The Potential of Information and Telecommunications Technologies for Rural Development. *Information Society, 9*(2), 145–156.

Mayo, J. K., Heald, G. R., & Klees, S. J. (1992). Commercial Satellite Telecommunications and National Development: Lessons from Peru. *Telecommunications Policy, 16*(1), 67–79.

Mbarika, V., Jensen, M., & Meso, P. (2002). International Perspectives: Cyberspace across Sub-Saharan Africa. *Communications of the ACM, 45*(12).

Meerbach, Gabriella. (1991). Senegal: Telephone Usage in a Third World Rural Area: With a Particular Emphasis on Women. *Intermedia, 19*(2), March/April, pp. 21–26.

Melody, W. H. (Ed.). (1997). *Telecom Reform: Principles, Policies and Regulatory Practices.* Lyngby, Denmark: Technical University of Denmark.

Meyer, A., Foote, D., & Smith, W. (1985). Communication Works Across Cultures: Hard Data on ORT. *Development Communication Report,* No. 51, Autumn.

Moyal, A. (1992). *Research on Domestic Telephone Use: Proceedings of a CIRCIT/Telecom Workshop.* Melbourne: Centre for International Research on Communication and Information Technologies (CIRCIT).

Mueller, M. (1997). *Universal Service: Competition, Interconnection, and Monopoly in the Making of the American Telephone System.* Cambridge, MA: MIT Press.

National Research Council, Board on Science and Technology for International Development. (1990). *Science and Technology Information Services and Systems for Africa.* Washington, DC: National Academy Press.

National Telecommunications and Information Administration. (1994). *20/20 Visions.* Washington, DC: U.S. Department of Commerce.

National Telecommunications and Information Administration. (1997). *Falling through the Net: Defining the Digital Divide*. Washington, DC: U.S. Department of Commerce.

National Telecommunications and Information Administration. (2001). *Networking the Land: Rural America in the Information Age*. Washington, DC: U.S. Department of Commerce.

National Telecommunications and Information Administration. (2002) *A Nation Online: How Americans are Expanding their Use of the Internet*. U.S. Department of Commerce.

Nice, P., & Johnson, W. (1999). *The Alaska Health Aide Program: A Tradition of Helping Ourselves*. Anchorage: University of Alaska. Center for Circumpolar Health Studies.

O Siochru, S. (1996). *Telecommunications and Universal Service: International Experience in the Context of South African Telecommunications Reform*. Ottawa: International Development Research Centre.

Okinawa Charter on the Global Information Society. www.g8.utoronto.ca/summit/2000okinawa/gis/html

OECD. (2001). *Understanding the Digital Divide*. Paris: OECD.

OECD. (2003). *Communications Outlook 2003*. Paris: OECD.

Page, M. (1977). *The Flying Doctor Story 1928–1978*. Adelaide: Rigby.

Parker, E. B. (1983). *Economic and Social Benefits of the Rural Electrification Administration (REA) Telephone Loan Program*. Geneva: International Telecommunication Union.

Parker, E. B., & Hudson, H. E. (1995). *Electronic Byways: State Policies for Rural Development Through Telecommunications* (2nd ed.). Washington, DC: Aspen Institute.

Pierce, W. B., & Jequier, N. (1983). *Telecommunications for Development*. Geneva: International Telecommunication Union.

Pool, I. de Sola (Ed.). (1977). *The Social Impact of the Telephone*. Cambridge, MA: MIT Press.

Portway, P. (1993). How Corporate America Trains by Telecommunications. *Communications News*, February, pp. 23–24.

Prahalad, C. K. (2005). *The Fortune at the Bottom of the Pyramid: Eradicating Poverty through Profits*. Upper Saddle River, NJ: Wharton School Publications.

Puma, M. J. et al. (2002). *The Integrated Studies of Educational Technology: A Formative Evaluation of the E-Rate Program*. Washington, DC: Urban Institute.

Qvortrup, L. (1989). The Nordic Telecottages. *Telecommunications Policy, 13*(2), 59–68.

Richardson, J. (1993). *Seven Steps to a Better Future for Rural Colorado*. Denver, CO: Advanced Technology Institute.

Robinson, S. S. "Rethinking Telecenters: Knowledge Demands, Marginal Markets, Microbanks, and Remittance Flows." www.isoc.org/oti/articles/0401/robinson.html

Rockoff, M. L. (1975). The Social Implications of Health Care Communication Systems. *IEEE Transactions on Communications*, Vol. Com-23, No. 10, October, pp. 1085–1088.

Rogers, E. (1995). *Diffusion of Innovations* (4th ed.). New York: Free Press.

Romero, S. (2001). Tight Bandwidth snarls Web Traffic in Middle East. *New York Times*, December 10.

Rosser, J. C., Jr., Gabriel, N., Herman, B., Murayama, M. (2001) Telementoring and Teleproctoring. *World Journal of Surgery, 25*(11), 1438–1448.

Saunders, R., Warford, J., & Wellenius, B. (1994). *Telecommunications and Economic Development* (2nd ed.). Baltimore: Johns Hopkins University Press.

Scharffenberger, G. (1999, September). Telecentre Evaluation Methods and Instruments: What Works and Why? *Proceedings of the IDRC Workshop on Telecentre Evaluation: A Global Perspective*. Far Hills, Quebec.

Schmandt, J., Williams, F., Wilson, R. H., & Strover, S. (1990). *Telecommunications and Rural Development: A Study of Business and Public Service Applications*. Austin: University of Texas at Austin.

Schramm, W. (1977). *Big Media, Little Media.* Beverly Hills, CA: Sage.

Schwankert, S. (1995). Dragons at the Gates. *Internet World,* November, p. 112.

Shapiro, C., & Varian, H. (1998). *Information Rules: A Strategic Guide to the Network Economy.* Boston: Harvard Business School Press.

SRI International. (1998). *Design of the WorLD Monitoring and Evaluation Component.* Menlo Park, CA: SRI International.

Stewart, A. (1994). Greasing the Wheels of Claims Processing. *America's Network. 1*(4), 37.

Taylor, J., & Williams, H. (1990). The Scottish Highlands and Islands Initiative: An Alternative Model for Economic Development. *Telecommunications Policy, 14*(3), 189–192.

Telecom Regulatory Authority of India. (2000). *Consultative Paper on Issues related to Universal Service Obligations.* New Delhi: TRAI, July 3.

Tietjen, K. (1989). *AID Rural Satellite Program: An Overview.* Washington, DC: Academy for Educational Development.

Tifft, J. D. (2002). *Summative Evaluation: Bulgaria Computer and Communication Center (PC3) Project.* Washington, DC: Academy for Educational Development.

United States Congress. (1993). *Hearing on Communication Benefits to Education and Finance before the Subcommittee on Communications of the Committee on Commerce, Science and Transportation.* Washington, DC: United States Senate, March 31, 1993.

United States Congress. (1996). *Telecommunications Act of 1996.* Public Law 104-104, February 8.

Waverman, L., & Meschi, M., & Fuss, M. (2005). The Impact of Telecoms on Economic Growth in Developing Countries. *In Africa: The Impact of Mobile Phones.* Vodaphone Policy Paper Series 2. www.vodaphone.com

Witherspoon, J. P., Johnstone, S. M., & Wasem, C. J. (1993). *Rural Telehealth: Telemedicine, Distance Education, and Informatics for Rural Health Care.* Boulder, CO: WICHE Publications.

WorLD Case Study Reports: Uganda. (1999). Washington, DC: World Bank, World Links for Development Program.

Wysocki Jr., B. (1997). Development Strategy: Close Information Gap. *The Wall Street Journal,* July 7.

Author Index

Subject Index

Milton Keynes UK
Ingram Content Group UK Ltd.
UKHW022108141024
449569UK00031B/1825